# Northwest Marine Weather

## From the Columbia River to Cape Scott

**Including Puget Sound, the San Juan and Gulf Islands,
and the Straits of Juan de Fuca,
Georgia, Johnstone, and Queen Charlotte**

*Jeff Renner*

THE MOUNTAINEERS

First edition: first printing 1993, second printing 1997



Published by The Mountaineers
1001 SW Klickitat Way, Suite 201, Seattle, Washington 98134

Published simultaneously in Great Britain by Cordee, 3a DeMontfort Street, Leicester, England, LE1 7HD

Manufactured in the United States of America

Edited by Don Graydon
Illustrations by Nick Gregoric
All photographs by the author
Cover design by Watson Graphics
Typesetting and layout by The Mountaineers Books

Cover photograph: Swiftsure race, Strait of Juan de Fuca
(© William W. Bacon III/AllStock)

Library of Congress Cataloging in Publication Data
Renner, Jeff.
Northwest marine weather : from the Columbia River to Cape Scott, including Puget Sound, the San Juan and Gulf Islands, and the Straits of Juan de Fuca, Georgia, Johnstone, and Queen Charlotte / Jeff Renner
p. cm.
Includes index.
ISBN 0-89886-376-7
1. Northwest, Pacific--Climate. 2. Maritime meterology--Northwest, Pacific. 3. Weather forecasting--Northwest, Pacific. I. Title.
QC984.N97R44 1993
551.69795--dc20 93-34608
CIP

♻ Printed on recycled paper

# Contents

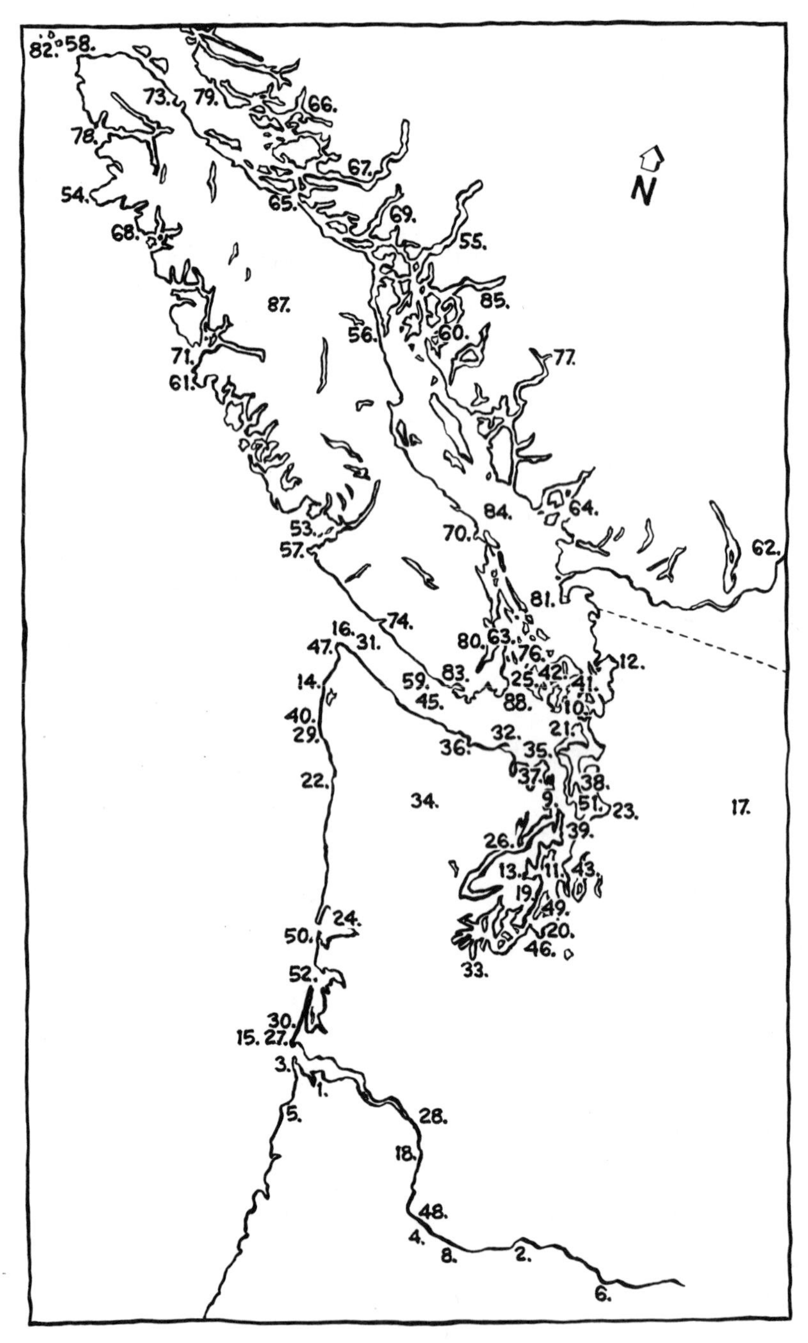


**Fig. 1.** *The Pacific Northwest*

**Oregon**

1. Astoria
2. Cascade Locks
3. Clatsop Spit
4. Portland
5. Seaside
6. The Dalles
7. Tillamook Head
8. Troutdale

**Washington**

9. Admiralty Inlet
10. Anacortes
11. Bainbridge Island
12. Bellingham
13. Bremerton
14. Cape Alava
15. Cape Disappointment
16. Cape Flattery
17. Cascade Mountains
18. Columbia River
19. Colvos Passage
20. Commencement Bay
21. Deception Pass
22. Destruction Island
23. Everett
24. Grays Harbor
25. Haro Strait
26. Hood Canal
27. Ilwaco
28. Kelso
29. La Push
30. Long Beach
31. Neah Bay
32. New Dungeness
33. Olympia
34. Olympic Mountains
35. Point Wilson
36. Port Angeles
37. Port Townsend
38. Possession Sound
39. Puget Sound
40. Quillayute
41. Rosario Strait
42. San Juan Islands
43. Seattle
44. Shelton
45. Strait of Juan de Fuca
46. Tacoma
47. Tatoosh Island
48. Vancouver
49. Vashon Island
50. Westport
51. Whidbey Island
52. Willapa Bay

**British Columbia**

53. Barkley Sound
54. Brooks Peninsula
55. Bute Inlet
56. Campbell River
57. Cape Beale
58. Cape Scott
59. Carmanah Point
60. Desolation Sound
61. Estevan Point
62. Fraser River
63. Gulf Islands
64. Howe Sound
65. Johnstone Strait
66. Kingcome Inlet
67. Knight Inlet
68. Kyuquot
69. Loughborough Inlet
70. Nanaimo
71. Nootka Sound
72. Porlier Pass
73. Port Hardy
74. Port Renfrew
75. Powell River
76. Prevost Passage
77. Princess Louisa Inlet
78. Quatsino Sound
79. Queen Charlotte Strait
80. Saanich Inlet
81. Sand Heads
82. Sartine Island
83. Sheringham Point
84. Strait of Georgia
85. Toba Inlet
86. Vancouver
87. Vancouver Island
88. Victoria

# Preface

When I was growing up, the television programs that fascinated me most were the undersea specials produced by Jacques Cousteau. Whatever license Cousteau may have taken with reality, his work stirred a profound fascination with the underwater environment and a strong desire to "set out to sea." Before I was old enough to earn my scuba certification or to take command of a boat, I spent every available moment in the water, until I was all prune-fingered and sniffly. Even then I often required considerable "guidance" to come out onto land and into dry clothes.

That fascination simply broadened as I matured into "a bigger boy capable of playing with bigger toys." A favorite quote is, "You cannot explore new waters without first setting out from shore." My enjoyment of new adventures has drawn me to leave the shore behind many times, setting out to sail, to kayak, and to explore underwater with scuba gear. I have been blessed to live in an area perhaps unmatched in the wonder and variety of its marine environment—the Pacific Northwest.

As science editor for Seattle's KING–TV, I was fortunate to have the support to do the first underwater film special documenting the marine life of this area with producer/photographer/dive partner Craig Johnston. Working on that special exposed us to some unforgettable creatures, including a first-hand encounter, in the water, with an orca. Those dives, as well as countless days spent with my wife, Sue, aboard our sailboat *Glissade* exploring from Puget Sound to Desolation Sound, weren't always under blue skies. I've had dive trips where I've stayed drier than when I never ventured from the tiller of our sailboat!

The days spent on, under, or above the water while sailing, kayaking, diving, and flying seaplanes exposed me to a wide range of Northwest weather phenomena. That hard-won practical knowledge has proven a useful partner to the formal training I received when earning my degree in atmospheric sciences at the University of Washington. As someone who has frequently been on the "consuming" end of weather forecasts, I part company with those colleagues who insist forecast users should be content to simply accept the forecasts prepared by meteorologists. Meteorology is an advancing, but as yet inexact, science. I know firsthand the limitations of such forecasts in the widely varied marine environment of the Northwest, how conditions can change rapidly, and the occasional need of the mariner to make as informed a decision as is possible immediately, without waiting for an updated forecast or advisory. As a meteorologist who has served as an expert witness (I always use the term "expert" loosely), I've seen too many sad accidents that resulted from a lack of basic weather knowledge, inadequate pre-trip planning, and insufficient vigilance.

The seeds of this book, then, were already firmly planted when I completed my previous book, *Northwest Mountain Weather.* Requests by reviewers and many pleasure boaters I've talked with to do a book on this region's marine weather gave the necessary impetus to complete this guide. Readers of *Northwest Mountain Weather* will recognize much of the material in the first two chapters, which cover the basic climatology of the Pacific Northwest and fundamentals of the way our planet's atmosphere operates. In the following three chapters, I distill the essentials of marine weather, emphasizing how they apply to sailing, cruising, and kayaking. In chapters 6 and 7, you'll find tips on selecting some important weather safety tools for your vessel or home, sources of weather information, and guidance on gathering the data and forecasts necessary for your planned trip. There are concrete outlines and examples of how to analyze that information, to make good "go" or "no-go" decisions. In chapter 8, organized in a decision-tree format, a series of checklists will assist you in assessing a wide variety of situations. The final chapter is essentially a field guide within a guide, covering the local peculiarities of each region within the Pacific Northwest. Use it as a planning aid to check the weather challenges most frequently found in the waters ahead, and refer to it when you have a question while under way.

If I were to sum up the purpose of this book, it would be to serve as a guide so you don't find yourself out on the water wishing you were home, or at home wishing you were out on the water. Now, for those cases where you find yourself at work wishing you were out on the water, there's nothing I can do about that! Here's to many fine days, and fair winds on the water.

*Jeff Renner*

# Acknowledgments

Writing a book is somewhat like raising a child; although there may be two biological parents to accept credit for the fine qualities of the offspring (and take responsibility for any problems), in truth there's an extended family, including relatives, friends, and teachers, responsible for moving the child from birth to maturity. So it is with this book. Special thanks go to my wife, Sue, for accepting hours of seeing only the back of my head as I worked on this manuscript and for accompanying me in weather fair and foul and water warm and cold on countless sailing and diving trips; to my son, Eric, for giving up time with Dad; and to my parents for their support: without them this book never would have been written!

Appreciation is also due the faculty of the Department of Atmospheric Sciences at the University of Washington, where I received my degree. The research of many, in particular Cliff Mass, Richard Reed, Peter Hobbs, and Mark Albright, has clarified many of the previously mysterious weather patterns of this region. Many of the tips contained in this book are derived from their research, from lectures attended during my student days, or from conversations. Every mariner owes these researchers a vote of thanks, as their work is behind both this volume and every successful forecast.

I would like to thank The Mountaineers Books and in particular my editor, Don Graydon, for their patient, diplomatic, and skilled guidance in moving this book from idea to published work. I also want to thank all the viewers who have called me at KING–TV with observations and questions that have helped refine my understanding of this region's weather. Finally, appreciation is due to all who have served as crew on this skipper's vessels, put up with me as crew on their boats, or accompanied me into the often murky depths as dive partners. It's been illuminating and fun, and I appreciate each friend who has accompanied me on such "field research trips." May we have many more!

# A Note About Safety

Safety is an important concern in all outdoor activities. No book can alert you to every hazard or anticipate the limitations of every reader. Therefore, the descriptions in this book are not representations that a particular day, place, or excursion will be safe for your party. When you engage in marine activities, you assume responsibility for your own safety. Under normal conditions, such excursions require the usual attention to wind, precipitation, sea state, the capabilities of your vessel and your party, and other factors. Keeping informed on current conditions and exercising common sense are keys to a safe, enjoyable outing.

*The Mountaineers*

CHAPTER 1

# Climate and Weather of the Pacific Northwest Coast

*My wife and I were enjoying our first vacation aboard* Glissade, *a beautiful Columbia 26 sailboat. We were cruising in the San Juans and had pulled anchor from Sucia Island, north of Orcas Island. A weak cold front had moved through overnight, and the light rain had given way to sunshine. The listless southerlies of the previous day had changed to a brisk northerly. It made for an exhilarating sail past the east shoreline of Orcas. As we turned to a westerly heading around Lawrence Point, the wind died. After all, we were close to the lee shore of Orcas, cruising through Rosario Strait. As we slipped through Obstruction Pass, prepared to enter East Sound, I noticed the water was becoming rougher. Guessing it was just a tide rip, I examined the chart in vanishing daylight for possible anchorages in Buck Bay, just outside the town of Olga. BANG! The semi-listless sails snapped full and heeled the boat so hard that the port rail almost touched water before Sue could even ask, "What was that!" After offering a few words of benediction, I mentally kicked myself, knowing I had ignored a clue that was warning me of strengthened winds roaring through East Sound. We had encountered gap winds, the result of the northerly wind being funneled through a relatively narrow gap in the terrain and accelerating as it was channeled down East Sound—just one of many examples of the subtleties and hazards of marine weather in the Pacific Northwest.*

The Pacific Northwest is a spectacular region born of the fire of volcanism and sculpted by water and ice. The 1980 eruption of Mount St. Helens and the occasional earthquakes that rattle this area serve as proof that this process of building and destruction continues, that the mountains of the Pacific Northwest remain geological adolescents. Understanding the geology of this region is essential to understanding its weather patterns, and understanding its weather patterns is essential to planning safe and enjoyable voyages on its waters.

The mountains of the Pacific Northwest, like those elsewhere, were created by geological forces acting deep beneath the earth's surface. Our

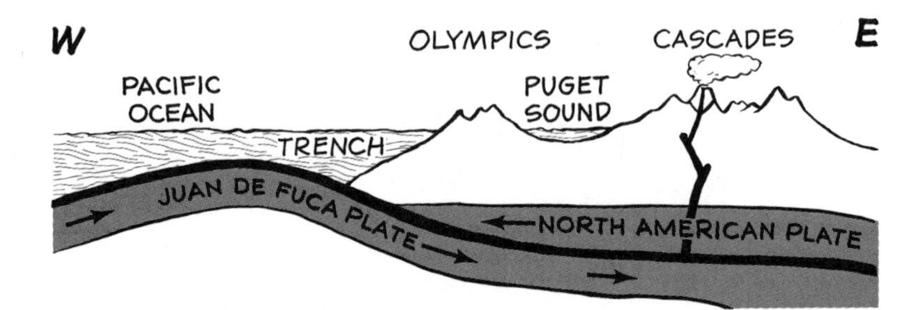


**Fig. 2.** *Mountain- and volcano-building in the Northwest*

planet's continents and ocean floors are really a collection of picture-puzzlelike pieces known as plates. These plates float on rock that is partially melted in the upper region of the earth's interior, called the mantle, and fully melted deeper within this layer. Temperatures in the mantle may range from 2,000 to 3,400 degrees Fahrenheit (1,100–1,900° C). This melted rock flows in currents, slow as cold molasses, that drive the massive plates, inch by inch.

In the Pacific Northwest, a plate making up part of the Pacific Ocean floor is colliding with and diving beneath the lighter plate that makes up our continent (fig. 2). Because the boundary between these colliding plates runs essentially north–south, the major mountain ranges thrust up by the collision also run essentially north–south.

Visitors flying into the Pacific Northwest on a clear day enjoy a spectacular sight: the blue waters of the Pacific giving way to a rugged coastline marked by the Coast Range of Oregon and Washington, interrupted by the broad Columbia River and then the Chehalis River Gap in southwestern Washington. The young Olympic Mountains rise to the north, separated from British Columbia's Vancouver Island Range by the Strait of Juan de Fuca, which is the shipping thoroughfare from the Pacific into Puget Sound to the south and Howe Sound to the north.

A large trough runs along the east side of these ranges, beginning with Oregon's Willamette Valley and continuing northward into the Puget Sound basin in Washington and then the Strait of Georgia in British Columbia. Farther east, the land rises again to form the Cascade Range, which extends from northern California through Oregon and Washington and finally merges with British Columbia's Coast Mountains. Two great valleys carve their way through the Cascades: the Columbia River Valley in Washington and Oregon and the Fraser River Valley in British Columbia.

These mountain ranges split Washington, Oregon, and British Columbia into distinctly different climate zones. These differences and the resulting variations in plant and animal life are consequences of the way the mountain ranges and principal bodies of water interact with the region's major weather patterns.

# Variations in Precipitation

Because most of the storms that move into the Pacific Northwest form over the Pacific Ocean and are directed eastward by winds high in the atmosphere, they end up on a collision course with these mountain ranges. As the moisture-laden air collides with the mountains, it's forced to rise. The resulting cooling triggers a massive growth spurt in the clouds and a large increase in precipitation along and near the mountains' western slopes. The process is similar to wringing out a sponge; the effect is certainly the same.

Even the most stately of sailing craft is dwarfed as it passes the towering stands of western hemlock, Douglas fir, and western red cedar which cover the moisture-rich western slopes of the coastal mountains. Shore explorations which extend beyond colorful tidepools brimming with anemones and sea stars may lead across the often spongy forest floor, an ideal habitat for the squishy banana slug that inspires either delight or disgust.

Just over the Cascades and Coast mountains, a startling transformation occurs. Skippers who trailer their boats east to Washington's Lake Chelan or cruise up the Columbia River past Oregon's Hood River find widely spaced, drought-tolerant lodgepole pines replacing the firs found to the west. The ground is often rock-hard, with sparse vegetation, and rattlesnakes may even be found sunning themselves on warm boulders.

The reason for the stark contrast is simple. Once over the mountain crest, the air, already relieved of some of its moisture, dries further as it descends the eastern slopes and warms. The result is a dramatic variation in precipitation, even over short distances (fig. 3). Quillayute Airport, near La Push on the northern Washington coast, averages 105 inches of rain each year (267 cm). Sequim, just over the Olympic Mountains along the Strait of Juan de Fuca, averages only 17 inches (43 cm)—less than 20 percent of what falls at the Quillayute Airport, which is only 64 miles away.

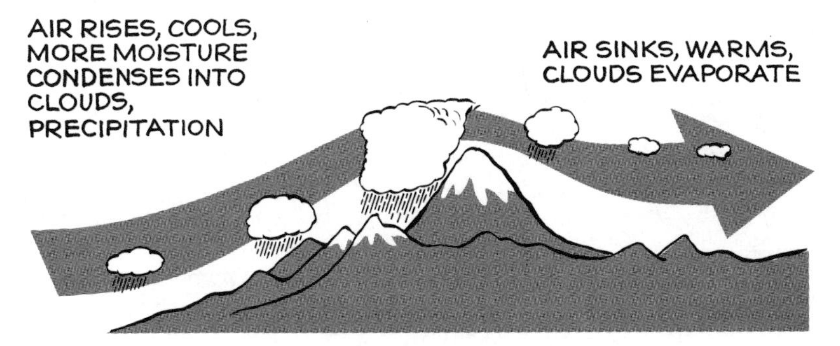
AIR RISES, COOLS, MORE MOISTURE CONDENSES INTO CLOUDS, PRECIPITATION
AIR SINKS, WARMS, CLOUDS EVAPORATE

**Fig. 3.** *Influence of mountains on precipitation*

An even more dramatic variation can be seen across the Washington Cascades. Stampede Pass, in the central Cascades, receives 91 inches (231 cm) of precipitation a year, while Yakima, to the east and in the shadow of the Cascades, averages only 8 inches (20 cm). Variations of this magnitude are unheard of in other regions of the United States and Canada but are common in the Pacific Northwest.

Forcing moist air up hills just 200 feet (60 m) higher than the surrounding terrain can double or even triple short-term precipitation (over a period of 12 hours or less)—and the Cascade crest ranges in elevation from 4,000 to 9,000 feet (about 1,200–2700 m). But large variations in precipitation amounts also occur between sites at the same elevation, such as at sea level, due to differences in adjoining slopes or exposure to prevailing winds.

## Precipitation in British Columbia

The Vancouver Island Range, which forms the backbone of British Columbia's Vancouver Island, produces strong contrasts in weather. The west coast of Vancouver Island, exposed to the full fury of incoming Pacific storms, is well known as a graveyard for ships. Reef-strewn Barkley Sound, a favorite destination of divers, is littered with the carcasses of ships that foundered as their crews sought refuge from hurricane-strength winds and downpours. Precipitation rates (fig. 4) increase rapidly upslope, exceeding 125 inches (317 cm) yearly on Mount Modeste and Mount Todd. The rate drops to less than 30 inches (76 cm) on some of the Gulf Islands to the east; it's no wonder this area is called the Sunshine Coast.

Farther east, across the Strait of Georgia and through Burrard Inlet, the city of Vancouver receives an average of 47 inches (119 cm) of precipitation each year. Precipitation amounts then increase rapidly with elevation to the east, with yearly averages of 95 inches (241 cm) near the crest of the Coast Mountains. Descending the eastern slopes of the range and out onto the Fraser Plateau, the average drops off rapidly, to as little as 16 inches (41 cm) a year.

## Precipitation in Washington

In Washington, variations in precipitation are equally impressive (fig. 5). The Coast Range boosts average annual precipitation to as much as 120 inches (305 cm) just east of Willapa Bay in southwestern Washington, with a rapid drop to only 40 inches (102 cm) farther east in Chehalis. By far the most dramatic variation, though, is seen in and around the Olympic Mountains. Hoh Head, along the rugged northern Washington coast, averages 90 inches (229 cm) a year. Mount Olympus, only 35 miles (56 km) to the east, averages 240 inches (610 cm) near its summit.

Moving east to the Puget Sound basin, Seattle receives an average of roughly 36 inches (about 90 cm) a year. Issaquah, a suburb in the foothills

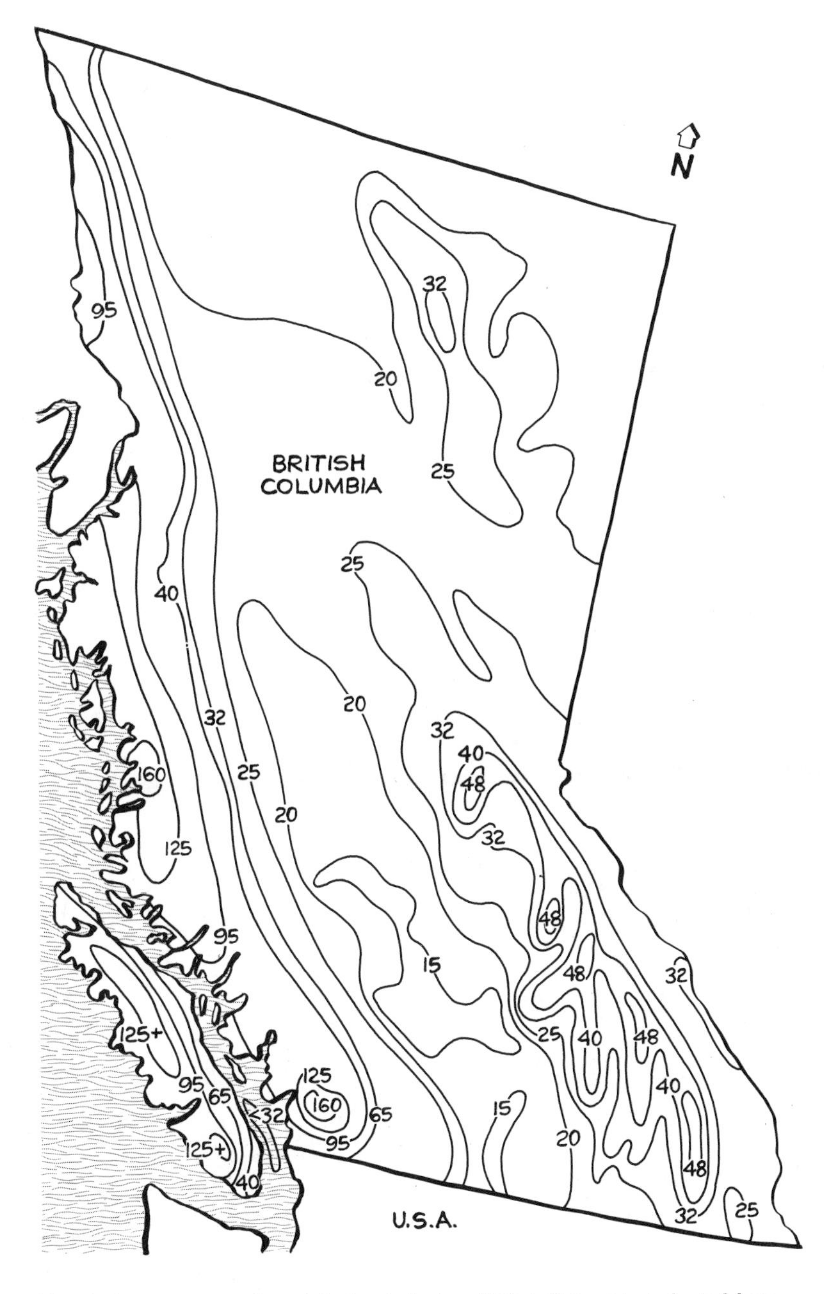


**Fig. 4.** *Average annual precipitation in inches, British Columbia (adapted from Environment Canada data)*

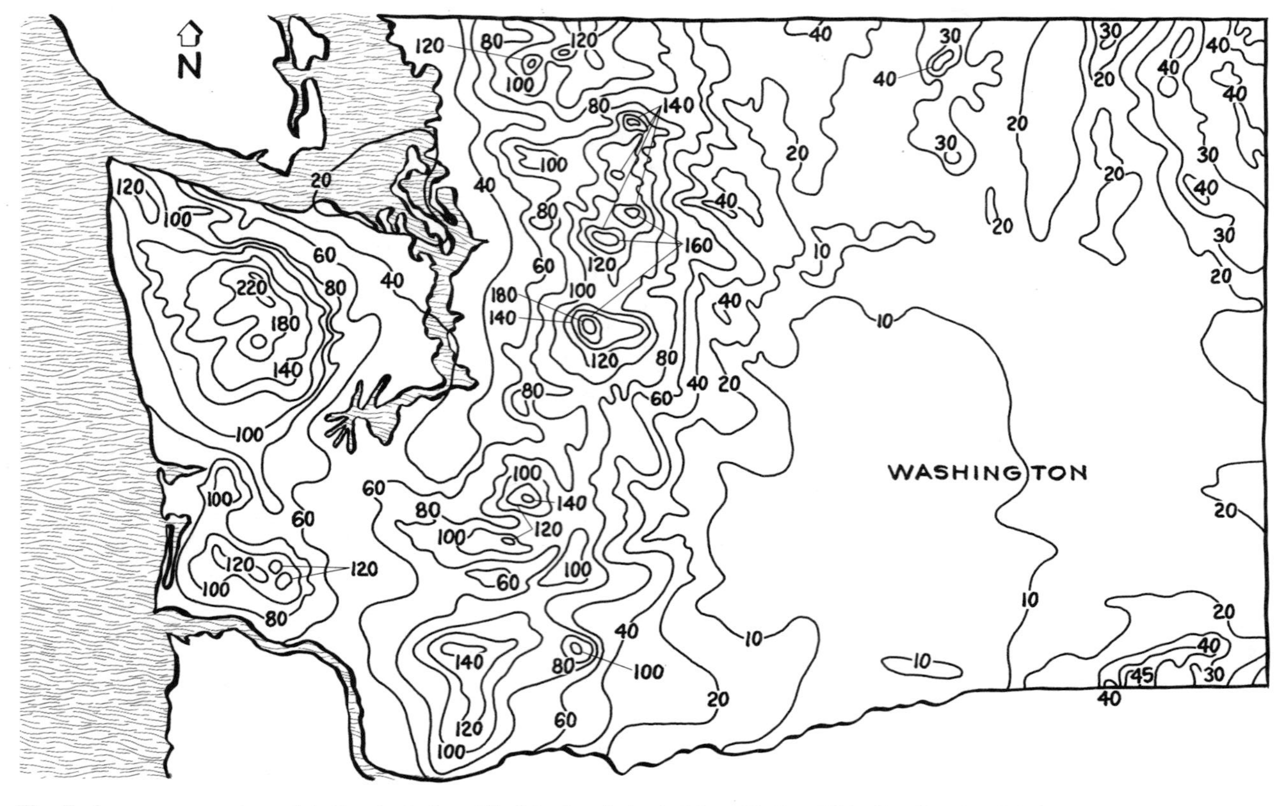


**Fig. 5.** *Average annual precipitation in inches, Washington (adapted from National Weather Service data)*

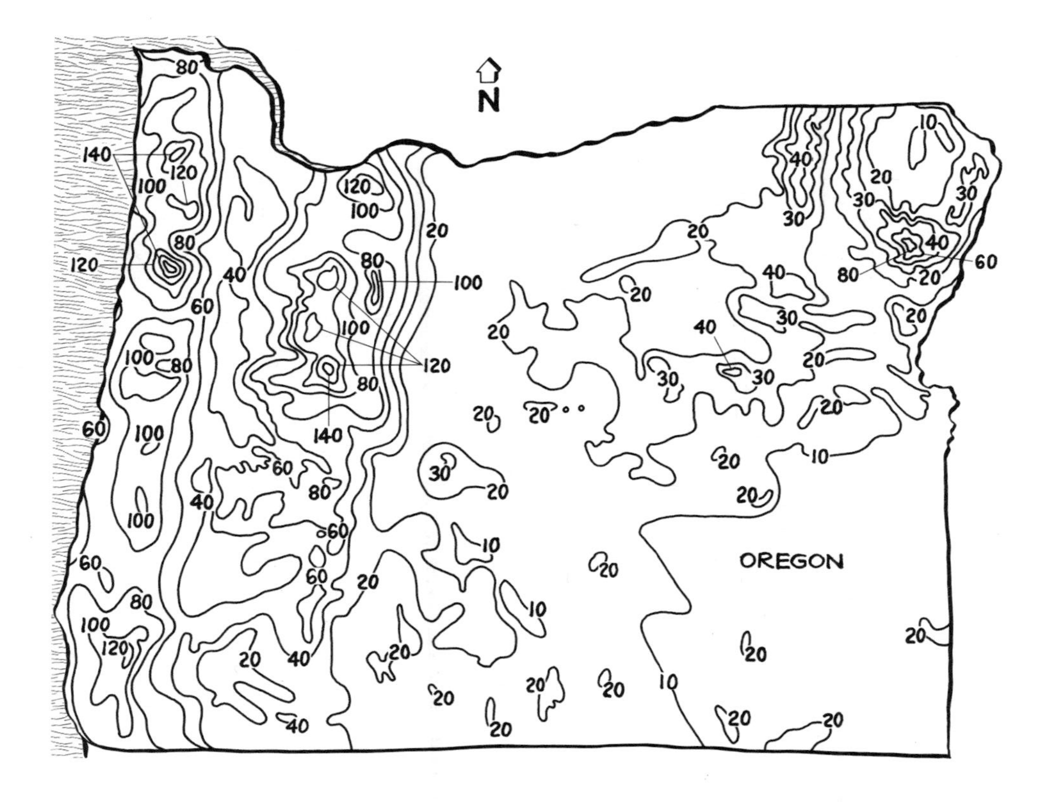


**Fig. 6.** *Average annual precipitation in inches, Oregon (adapted from National Weather Service data)*

of the Cascades, receives 50 inches (127 cm). Yearly averages of more than 100 inches are not uncommon along the Cascades to the east, with as much as 190 inches (483 cm) near Monte Cristo.

The dry east side of the Washington Cascades is truly arid, with much of the area receiving annual rainfall of 10 inches (25 cm) or less.

## Precipitation in Oregon

Oregon also experiences startling variations in precipitation (fig. 6). Annual rates increase abruptly just east of Lincoln City along the coast, from 70 inches (178 cm) along the beaches to as much as 200 inches (508 cm) near the summits of the Coast Range—then drop abruptly to just 40 inches (102 cm) in Salem, sheltered in the Willamette Valley.

Suburban Hillsboro, to the west of Portland, receives an average of 38 inches (97 cm) per year. Troutdale, to the east of Portland, averages 48 inches (122 cm). On Mount Hood, that spectacular volcano and storm-maker to the east, precipitation averages soar to 130 inches (330 cm) per year. But once over the Cascades, amounts decline rapidly. The Dalles, along the Columbia River, averages less than 15 inches (38 cm).

# Variations in Temperatures

The Cascades of Washington and Oregon, and British Columbia's Coast Mountains, create more than wet and dry zones; they also play a major role in producing variations in temperature. The mountains tend to confine the moist ocean air to the west, which has a moderating effect on climate.

During the summer months, the mountains' role as a barrier to moist ocean air permits temperatures from the Cascade crest eastward to far exceed those to the west. The average July daytime high temperature in Seattle is 75 degrees Fahrenheit (24° C) while in Yakima it's 88 (31° C).

From October through March, the Cascades and British Columbia's Coast Mountains usually deflect to the east the bitterly cold arctic outbreaks from Alaska, the Yukon, and the Northwest Territories. The average January overnight low temperature in Seattle, for example, is a relatively mild 34 degrees Fahrenheit (1° C), while in Yakima it's 18 (–8° C).

# Different Storm Tracks, Different Weather

Because the Pacific is the birthplace of most of the storms that hit this region, forecasters pay close attention to the *source region* of the weather systems that develop. Weather systems moving from one area of the Pacific may produce prolonged, heavy rains that soak regions both east and west of the major mountain ranges; those from another area might generate intermittent showers confined to the west side of mountainous peninsulas or islands. Know where a storm was formed, and the path it has traveled (its storm track), and you have a quick way to assess its likely impact and to choose the destination offering the best weather for cruising, sailing, paddling, or fishing.

## From South–Southwest to Southwest

The Pineapple Express, true to its name, tends to direct warm and very humid air up from the subtropics or even the tropics (fig. 7). We can often see a band of clouds stretching from the Hawaiian Islands to

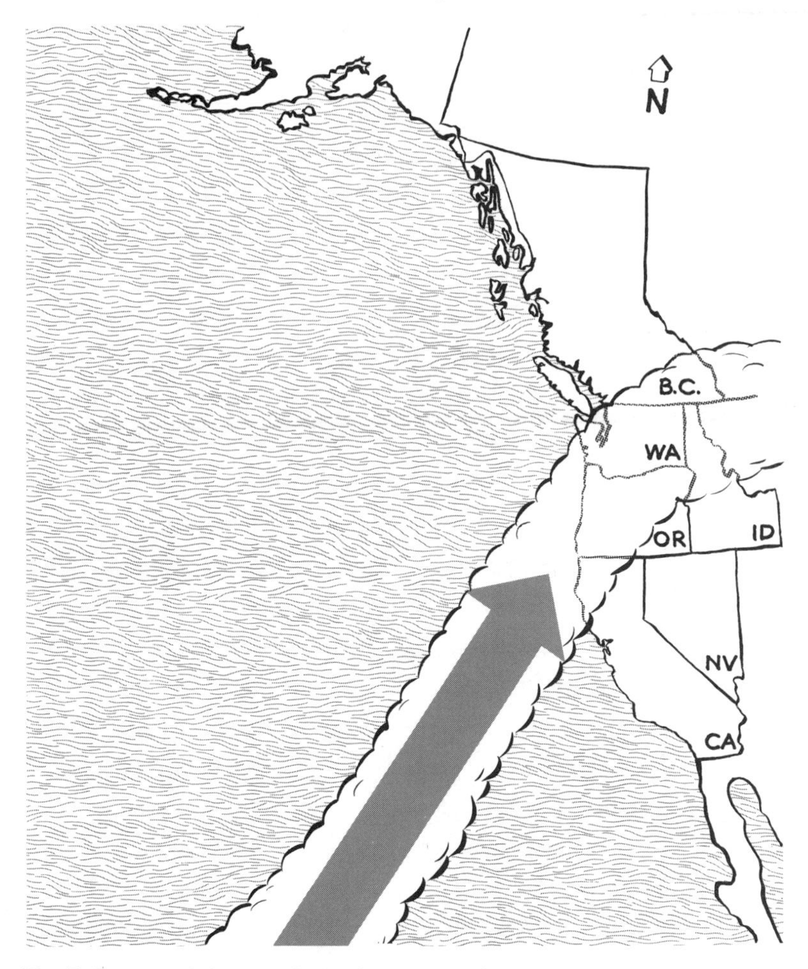


**Fig. 7.** *Storm track from south–southwest to southwest*

Vancouver Island in satellite pictures. Because the temperature and moisture content of this air is very high, we get rain when it collides with the much cooler air in the Pacific Northwest and is lifted.

The Pineapple Express often brings 1 to 2 inches (2.5–5 cm) to places like Puget Sound and the Willamette Valley, and 3 to 4 inches (7.5–10 cm) to the Oregon, Washington, and British Columbia coasts. This storm track tends to persist for several days as new disturbances develop and ripple northeastward along the band of clouds.

The very mild air moving up from the tropics usually raises the freezing level to 8,000 feet (about 2,400 m) or higher. Snow melts rapidly in the mountains, often leading to serious flooding all the way down to sea level, particularly if coupled with unusually high tides. The air is usually fairly stable, so thundershowers are less common.

## From Southwest to West

A more common storm track is from the southwest to west (fig. 8). The temperature and moisture content of the air are usually lower; the result is less-intense rainfall (although not necessarily light). The rain isn't as persistent as with the Pineapple Express.

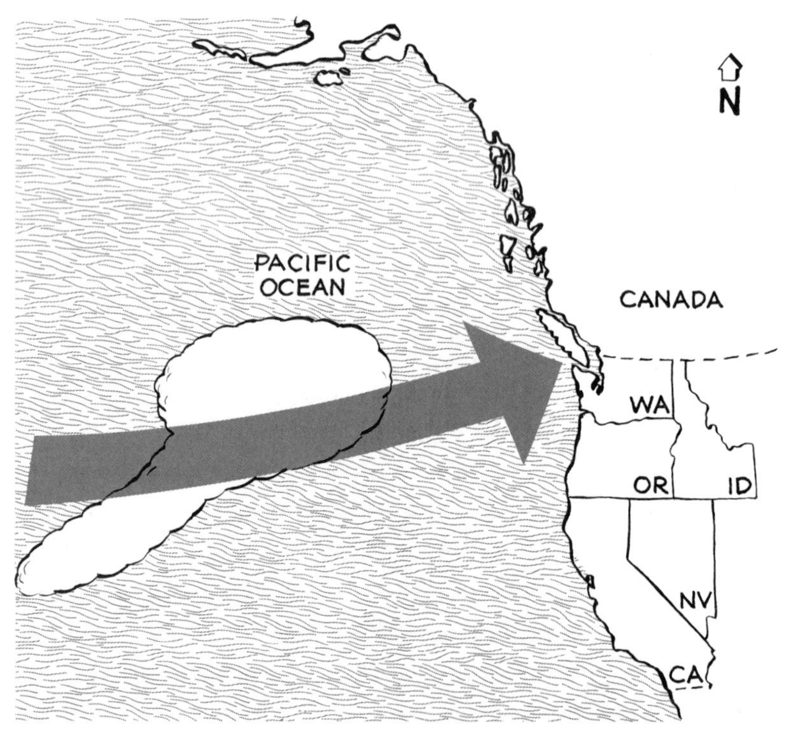


**Fig. 8.** *Storm track from southwest to west*

Although the prevailing westerlies can fire off one disturbance after another at us, there is usually a break between them—sometimes lasting only a few hours, occasionally a full day. Cooler air moves in after each disturbance, hence a better chance of thundershowers, especially along the windward sides of peninsulas and islands, where air is given an additional thrust upward by the terrain.

If cold air has an icy grip on the Northwest prior to the arrival of a system in this storm track, the precipitation may start out as snow but gradually change to rain. Freezing levels associated with this pattern vary from season to season, from 3,000–5,000 feet (about 900–1,500 m) in winter to 5,000–7,000 feet (about 1,500–2,100 m) in autumn and spring.

## From West to Northwest

Precipitation from this storm track (fig. 9) usually doesn't last long, but it can come as a rude surprise. When the storm track is generally from

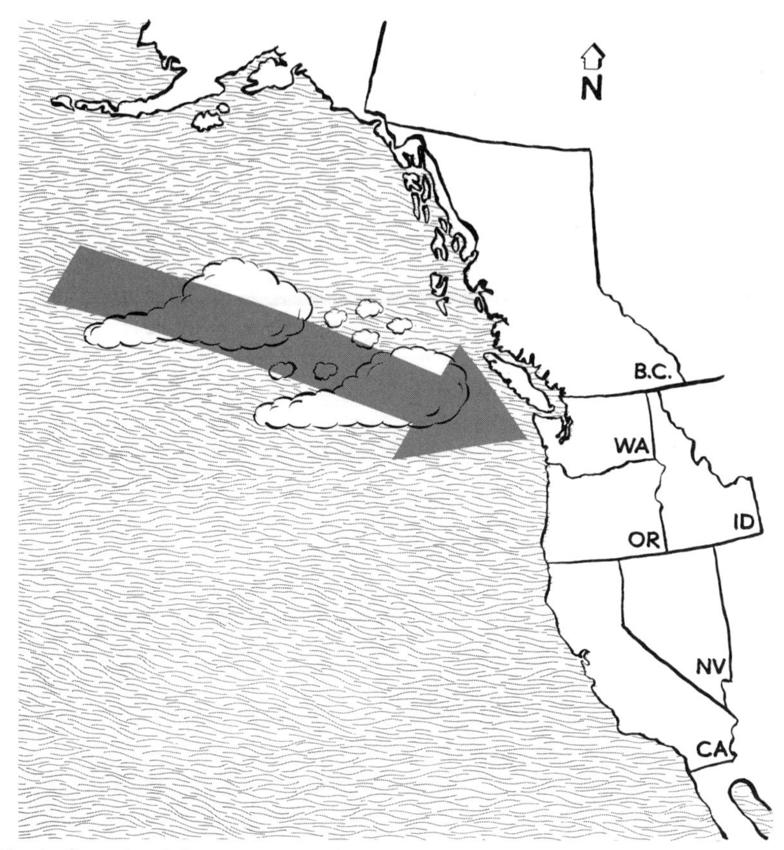


**Fig. 9.** *Storm track from west to northwest*

the northwest, weather systems move through quickly, with fairly rapid clearing. However, the contrast between temperatures before and after frontal passage in this pattern can lead to thundershowers after frontal passage. Expect freezing levels from just above sea level to 3,000 feet (about 900 m) in winter, to 3,000–5,000 feet (about 900–1,500 m) in autumn and spring.

## From North

A storm track from the north (figures 10 and 11) is this region's most frequent producer of snow along the southwestern British Columbia coast, the Puget Sound lowlands, and the northern Willamette Valley.

Cold air within a vast dome of high pressure leaks out from the interior

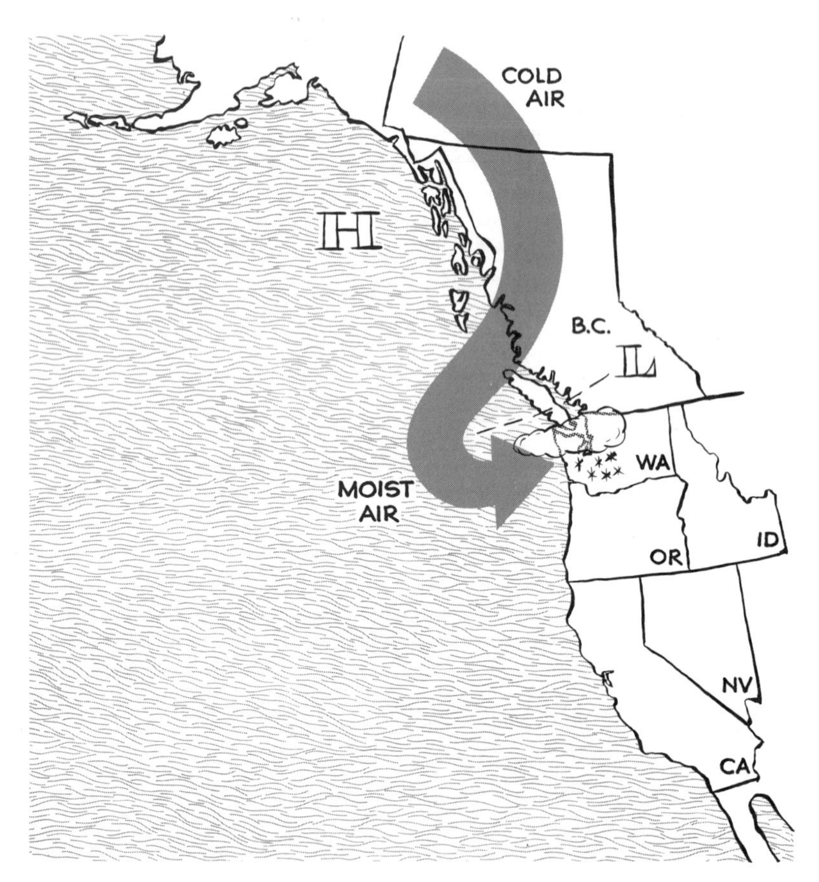


Fig. 10. *Northerly storm track*

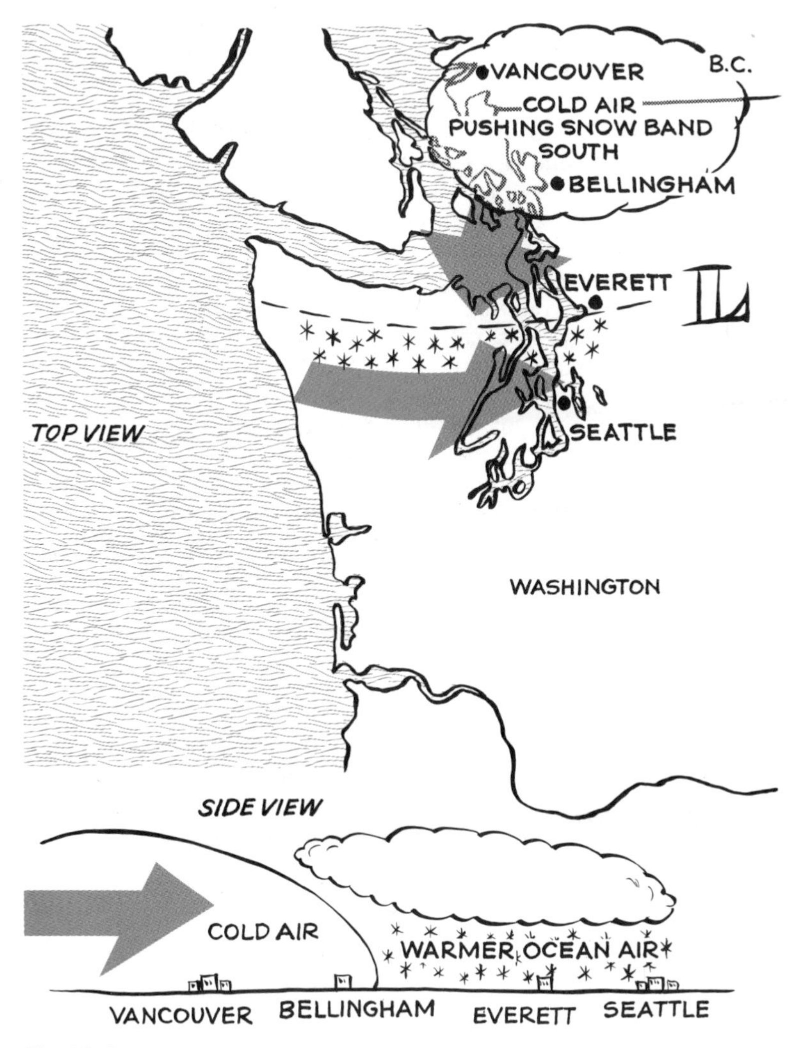


**Fig. 11.** *Snow pattern with northerly storm track*

of British Columbia through passes or river valleys (the Fraser River Valley is a favorite). This cold air circulates over the Pacific just long enough to pick up moisture, but not long enough to warm the air so that rain could be produced instead of snow. As the low that develops drops southward along the coast, snow makes its way from Vancouver to Bellingham to Everett to Seattle—sometimes all the way to Portland. This is the rare pattern where more snow occasionally falls in the cities and lowlands than in the mountains. This pattern, however, is not terribly long-lived, usually lasting a day at most.

## Latitude and Weather in the Northwest

The jet stream, that high and fast-moving river of air that directs storm tracks, is strongest and farthest south—and thus nearest to the Pacific Northwest—in the winter months. Therefore, that's when the Northwest receives most of its precipitation. Seattle and Portland receive less than 10 percent of their annual precipitation during the summer months of June, July, and August. Knowing this pattern can assist you in picking the best time of year for your marine vacation.

The retreat of the jet stream to more northerly latitudes during the summer doesn't benefit all areas equally. Although the jet stream tends to remain over southern Alaska and northern British Columbia, weak disturbances still move far enough south to make sailing or cruising around Vancouver Island, the Gulf Islands, or the San Juan Islands markedly cloudier and wetter than in south Puget Sound or the Columbia River. This tendency is especially strong along the west coast of Vancouver Island and up into the Queen Charlotte Islands. (The San Juan and Gulf islands are sheltered somewhat by the Olympic Mountains and the Vancouver Island Range.)

The climatic diversity is what makes our region so special and enjoying it on the water such a delight. It's all a consequence of the geological forces that continue to shape the Pacific Northwest and of the region's proximity to the weather factory of the Pacific Ocean. The result is not one climate zone but several. Visualizing the big picture is important to understanding the more subtle details of Pacific Northwest weather and to using that understanding to make trips on the water enjoyable and safe.

CHAPTER 2

# What Makes It Wet

*My phone rang a little after two in the afternoon. I had been studying the charts in my forecast office on an unseasonably warm and humid spring day. A vigorous cold front was swinging southeastward from the chilly Gulf of Alaska and was now not more than fifty miles off the Washington and British Columbia coasts. Weather observations from Tofino, Quillayute, and Hoquiam all reported towering dark cumulus clouds offshore, and an automated weather buoy showed rapid falls in pressure. The call was from a sailing friend who was planning to compete in a race on Lake Washington that evening but was concerned about the erratic movement of his barometer. "Go another night," I said. "This looks like an unusually strong cold front, the kind that could trigger thundershowers and possibly even a squall line in advance. Next time, pick a night I'm not working so I can go along!" Later that afternoon, NOAA weather radio was also cautioning mariners about the possibility of fast-moving thunderstorms and gusty winds. My friend stayed home that evening even though the race went on as planned. Skies over Lake Washington continued a hazy blue, but static electricity was interfering with AM radio reception, and tall cumulus clouds could be seen to the southwest. Thirty minutes after the race began, a squall line ripped across the lake, knocking down three sailboats, damaging others in their slips, and injuring several skippers and their crews.*

A little attention to television or NOAA weather forecasts, shipboard barometers, or the ominous cumulus clouds moving in from the southwest could have prevented the injuries and damage in this race on Lake Washington. The plummeting barometer and the approach of towering clouds both signaled the approach of a dangerous squall line.

At times, weather systems in the Northwest cover the entire region and your choice in deciding whether to go boating is a simple one: stay home. More frequently, patterns in this region are subtle and localized. While deciphering the clues can be difficult, such patterns offer sailors, cruisers, and kayakers more options than a simple go or no-go. As a step toward understanding these patterns, let's begin with some basics.

# A Little Basic Meteorology for Mariners

## Temperature

The sun does far more than simply illuminate our home planet. It is the engine that drives our atmosphere, providing the heating that, along with other factors, creates the temperature variations that are ultimately responsible for wind, rain, thunder, lightning—everything we call weather.

The earth's location—93 million miles (nearly 150 million km) from the sun—is what makes life as we know it possible. Venus, with an orbit closer to the sun, experiences average surface temperatures of roughly 800 degrees Fahrenheit (about 425° C), while the more distant Mars averages 81 degrees below zero (–63° C). Imagine the foul-weather gear needed for a visit to either of our neighboring planets!

Proximity to the sun is only one factor. The intensity of the sun's radiation varies across the earth's surface. Given a choice between an early spring sail in the San Juan Islands in Washington or the A-B-C islands (Admiralty, Baranof, and Chichagof) in southeastern Alaska, for instance, a sailor with limited tolerance for cold temperatures will likely choose the San Juans. That's because the San Juan Islands are closer to the equator; the sun will be more directly overhead at noon, and therefore the heating from the sun will be more intense.

This relationship between heating from the sun and the angle of the sun above the horizon also explains why summer is warmer than winter: the sun is more directly overhead. You can see how this works by shining a flashlight on this page, first from directly overhead, then at an angle. The beam of light shining from directly above the page has a smaller area to illuminate and heat than a beam striking the surface at an angle. The smaller the area illuminated by the flashlight (or the sun), the more intense the heating.

Given more intense sunlight at the equator than at the poles, the temperature differences come as little surprise. But extremes in temperature, large as they may be, are controlled by the movement of air. Differences in air temperature lead to air movement, which prevents runaway heating or cooling.

## Air Pressure

Anyone who's hoisted a sail knows that air moves sideways (except during a race, when it usually becomes dead calm). But air also rises and descends, movements that can generate or dissipate clouds.

When air rises due to heating, it's as if it were shedding extra pounds. Air has weight, and just as the reading on our bathroom scale drops when we lose weight, the reading on a barometer, which measures air pressure, falls when some of the air moves up and away.

Just as heating air makes it rise, cooling air makes it sink. Because cold air is more dense than warm air, it tends to find its way to the bottom of the atmosphere; that is, to the ground or water. Cold air tends to collect in sheltered inlets, making them a chilly anchorage on cold, windless nights.

To summarize, when cold air sinks, the air pressure increases; when warm air rises, the air pressure decreases. These pressure differences, the result of temperature differences, produce moving air. We call this moving air wind. Air generally moves from an area of high pressure to one of low pressure (fig. 12).

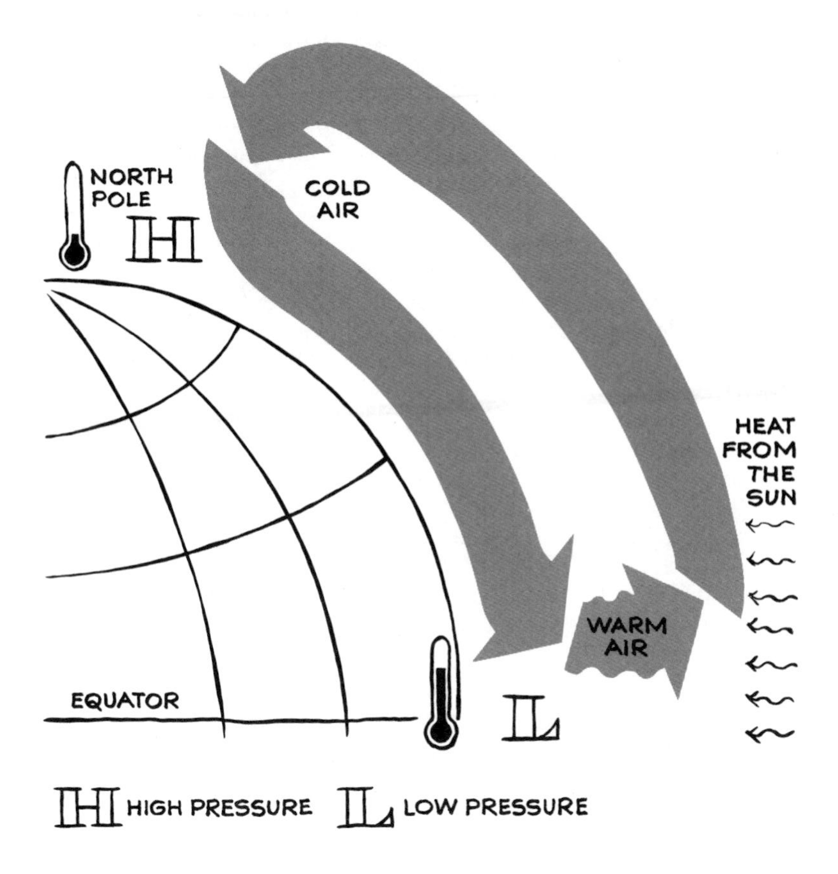


**Fig. 12.** *Relationship between solar heating and air pressure*

## Cloud Formation

Air moving from high to low pressure carries moisture with it. As that air cools, either from rising or moving horizontally over a colder surface, the moisture condenses into clouds or fog. This occurs because as air cools, its capacity to hold water vapor is reduced. For example, air at 98.6 degrees Fahrenheit (37° C), our body temperature, is capable of holding roughly thirteen times as much water vapor as it can at 30 degrees (–1° C).

When that moisture-laden air is cooled, its capacity to hold water vapor is rapidly reduced. Not all of the water vapor will "fit," and that which "spills out" condenses into a cloud of water droplets.

When air cannot hold any additional water vapor, we say it is saturated. Meteorologists call that saturation point the *dew point.* The dew point is simply the temperature at which the air will become saturated with moisture as the air cools, and clouds will usually form. We encounter a similar effect when we see our breath on a cold day. As we inhale, our body warms the air to approximately 98.6 degrees Fahrenheit and adds moisture. As we exhale, that warm, moist air is cooled by the colder air around us, leading to condensation of the water vapor (a gas) into water droplets (a liquid).

The dew point is always equal to or cooler than the air temperature, never warmer. When the air temperature cools to the dew point, water vapor condenses into water droplets, and clouds or fog form. No fog will form when the temperature is significantly higher than the dew point.

The comparison of how much water vapor the air is holding with how much it *could* hold is called *relative humidity*. Think of it as a measure of how saturated the air is. Relative humidity is usually expressed as a percentage; 75 percent relative humidity, for example, means the air contains three-quarters of the water vapor it's capable of holding.

The process of cooling and condensation operates on a large scale in the atmosphere as air moves from high-pressure into low-pressure systems and is lifted. The result is, at times, a weather system that covers the entire Northwest, though much of what we encounter on and off the water is more localized. Let's look now at large weather systems.

# Large-Scale Weather Systems

Experienced sailors and cruisers know that precipitation in the Pacific Northwest isn't always light and often encompasses the entire region in a soggy embrace. The cause? Large-scale weather systems, usually moving in from the Pacific.

The Gulf of Alaska produces most of the storms that batter the Pacific Northwest. This is a natural consequence of its latitude and geography.

The gulf is the battleground—between cold, dry air moving south from the Arctic and warm, moist air moving north from the subtropics and mid-latitudes—that sparks many of the storms that hamper our recreational trips on the water.

Because polar and arctic air is colder and therefore more dense than air farther south, it sinks. The zone where it sinks and "piles up" is a region of *high pressure.* As the air sinks and its pressure increases, its temperature also increases. The effect is similar to what happens to football players caught at the bottom of a pile. The players on the bottom get squeezed the most, and their temperature (and possibly their temperament) heats up. In the atmosphere, this warming within a high tends to evaporate the little moisture present in cold polar and arctic air. This is why the Arctic receives very little precipitation and why it is classified as a desert; not all deserts are covered by sand!

Barrow, for example, on the north slope of Alaska, receives an average of only 28 inches (71 cm) of snow each year, in contrast to Fairbanks, which receives 66 (168 cm), and Juneau, which averages 105 (267 cm). Each town is progressively farther south.

If our planet didn't rotate, this cold air would just continue to slide southward to the equator. Intense solar heating near the equator forces air to rise, creating a region of *low pressure* that rings the globe. Because air

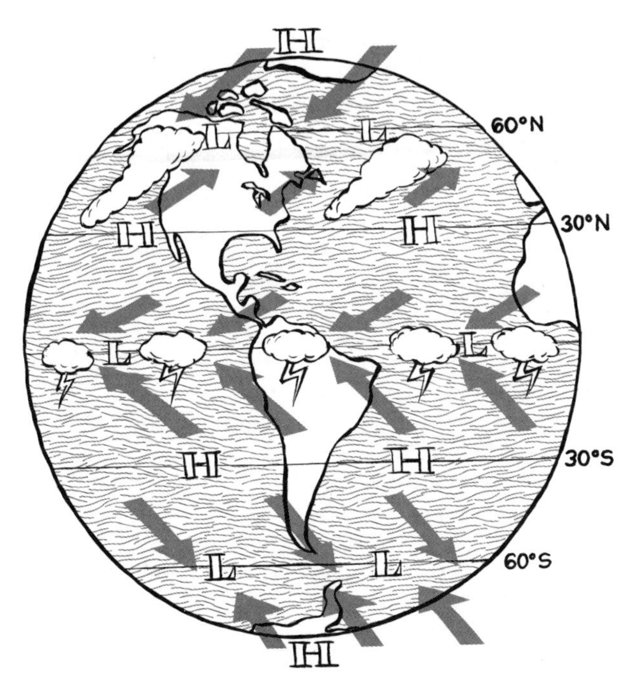


**Fig. 13.** *Circulation patterns*

within this band rises, it also cools, which tends to condense water vapor into liquid water droplets that form clouds, just as your breath condenses on a cold day. Satellite pictures show a series of thunderstorms marking this equatorial low-pressure zone, which is called the Intertropical Convergence Zone. It's a very wet area: more rain can fall in a single day within this zone than during an entire month in Washington, Oregon, or British Columbia.

But the air sinking and moving south from the pole and that rising from the equator don't form a simple loop moving from north to south and back again. The rotation of the earth is responsible for deflecting this air, creating a considerably more complicated circulation of air over our planet (fig. 13).

Some of the air rising from the equator descends over the subtropics. This sinking air creates a region of high pressure. As it sinks within this high, the earth's rotation produces a force known as the *Coriolis force.* The combination of the Coriolis force and the effect of friction from air moving over land and water causes the air sinking within the high-pressure system to rotate in a clockwise direction in the northern hemisphere and counterclockwise in the southern hemisphere.

## Fronts

Some of the air that sinks and spreads outward from these subtropical highs picks up moisture from the oceans and moves north, eventually meeting the cold, dry air spreading southward from the pole. The boundary between these two very different types of air masses is called the *polar front,* and it rings the globe in both hemispheres. Although the polar front moves north and south, its position averages between 50 and 60 degrees north and south latitude.

When this boundary between different air masses doesn't move, it's also called a *stationary front.* In the Gulf of Alaska and elsewhere, it serves as a nursery for the development of storms.

During the autumn and winter months, the air moving south from the Arctic toward the Gulf of Alaska can be as cold as 40 or 50 degrees below zero Fahrenheit (–40 to –45° C). The temperature of the air over the gulf is moderated by water's capacity to absorb and retain heat. Air temperatures there may be 30 to 40 degrees above zero (about –1 to +4° C), yielding an impressive contrast along this polar front of as much as 90 degrees Fahrenheit.

Because of this great contrast in temperatures, the polar front is especially strong in and around the Gulf of Alaska. But it rarely remains stationary there or anyplace else. Imbalances caused by the rotation of the earth and the differing influences of land, sea, ice, and mountains allow the cold, dry, dense air from the north to slide south, forcing some of the warm air to rise. The zone where the cold air is replacing the warm air is referred to as a *cold front* (fig. 14).

Conversely, farther east, warm air is forced to glide up and over the cooler air near the surface. This zone where warm air is gradually replacing cooler air is referred to as a *warm front* (fig. 14). This "wave" or bend on the stationary front may develop into a low-pressure system, with air circulating counterclockwise around the low (the opposite direction of air moving around a high)—again a consequence of the earth's rotation and friction.

When such low-pressure systems develop in the Gulf of Alaska, the counterclockwise circulation of air around the low draws warm, moist air northward from over the Pacific Ocean and very cold air southward from the interior of Alaska.

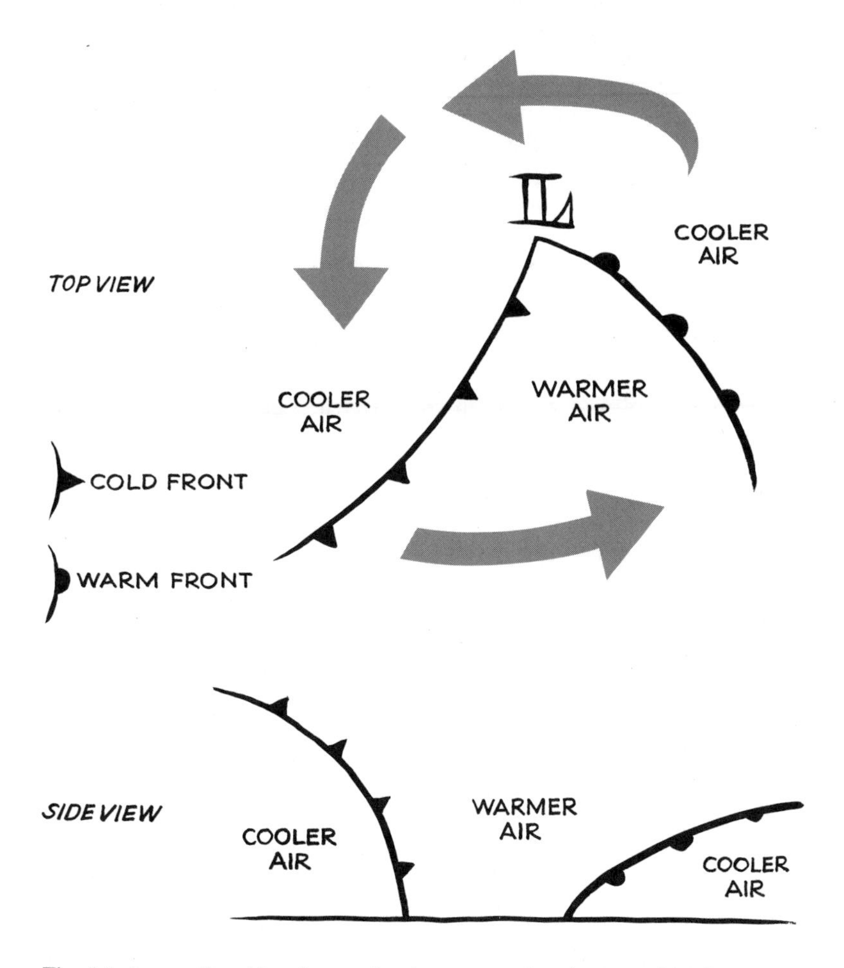


**Fig. 14.** *Low with cold and warm fronts:* top, *overview;* bottom, *side view*

# Clouds

Different types of clouds develop along cold and warm fronts, and it's the differences in their shapes that can provide the field forecaster with valuable clues to coming changes in the weather. Variations in cloud shapes along warm and cold fronts mirror differences in the physical processes that are taking place along those fronts. (See the cloud identification chart in Appendix 2.)

## Warm-Front Clouds

If a low-pressure system is moving across the Pacific, the first clouds visible in the Pacific Northwest are usually those associated with the warm front. Because the warmer air moving inland is less dense than the cooler air it's replacing, it is lifted, gradually sliding up and over the cooler air at the earth's surface. As that warm, moist air from the Pacific rises, it cools. If that rising air cools to its dew point, the water vapor in the air mass will either condense into water droplets or *sublime* into ice crystals.

Sublimation is the atmosphere's shortcut, transforming water vapor directly into ice crystals without it first condensing into liquid water. The result is determined by the temperatures aloft. Because the advancing warm air usually rises very high, 20,000 feet (6100 m) or more, the temperatures are well below freezing, certainly cold enough to sublime the water vapor into ice crystals.

### *Cirrus Clouds*

The first clouds we see as the warm front advances are the fibrous high ones called *cirrus clouds.* They are thin, often less than 1,000 feet (about 300 m) thick. The clouds' ice crystals or water droplets act as miniature prisms, bending and splitting sunlight or moonlight into its component colors. The result is the halo often seen ringing the sun or moon. Such halos are very wide and change in color from red at their inner ring to yellow to green to blue; they usually precede precipitation by 24 to 48 hours.

### *Lenticular Clouds*

Another type of cloud is created by moist air moving up and over a mountain. We see it frequently in the Northwest, especially above volcanic peaks such as Mount Rainier or Mount Hood. It's called a *lenticular cloud;* that is, a cloud shaped like a lens (fig. 15). Such clouds are often hints that a weather disturbance is nearby and that a warm front may be approaching.

Lenticular clouds are formed when moisture high in the atmosphere is deflected upward when it runs into a major peak. As that moist air rises, it cools sufficiently to condense into a cloud. But as it passes over the peak and begins descending, it warms, and the water droplets or ice crystals that make up the lenticular cloud evaporate. Although lenticulars appear to be stationary (they've often been mistaken for UFOs), they are continually dissipating on the leeward edge. If followed by cirrus and eventually stratus clouds, such lenticulars often give the observant mariner 24 to 48 hours' notice of approaching precipitation.

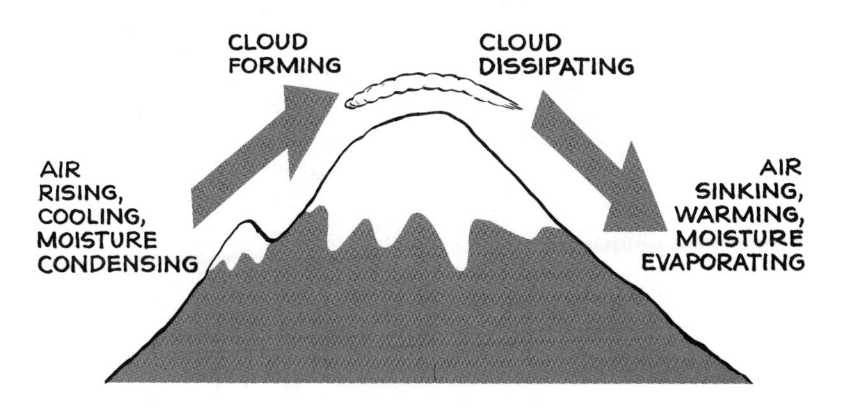


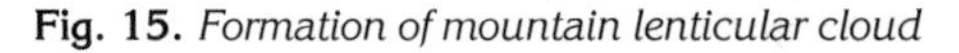
**Fig. 15.** *Formation of mountain lenticular cloud*

### *Lee-Wave Clouds*

The up-and-down motion of air traveling over a mountain peak or range occasionally continues to the downwind side, much like the ripples produced by a rock thrown into a pond. This results in a series of clouds forming and dissipating as the air rises and falls in a wavelike fashion.

Such clouds are called *lee-wave clouds* and can stretch for hundreds of miles downwind of the mountain barrier. Because these clouds form to the lee side of mountain ranges, they are most frequently found to the east of the Cascades in the Northwest. When oriented properly to the sun, wave clouds are often tinted in beautiful iridescent colors.

### *Stratus Clouds*

As the warm front advances, the boundary between the warm air and the cooler air below gradually lowers closer to the ground. More moisture is available at the warmer temperatures found closer to sea level, and thicker clouds form: sheetlike *altostratus clouds,* ranging from 20,000 to as little as 6,000 feet (about 6,100–1,800 m) above sea level, and thick, blanketlike layers of *stratus clouds,* which can extend right down to sea level. Stratus clouds can range in thickness from a few hundred to several thousand feet, while altostratus clouds usually are less than 1,000 feet thick.

Another circular "rainbow" of sorts often rings the sun or moon through altostratus. This is called a *corona,* and it hugs the sun or moon much more closely than a halo. For that reason, while halos tend to indicate that precipitation is at least 24 hours away, coronas suggest imminent precipitation, usually within the next 12 hours.

The flatness of these stratiform clouds is a consequence of what meteorologists call *stability*. Stability is simply the resistance of air to some force that is attempting to push it upward. Stable air tends to spread out in a layer; unstable air tends to balloon upward, like the bubbles in a pot of boiling water. When air cools very slowly with increasing altitude, or when warm air actually overlies cooler air near the surface, as in a warm front—or in an extreme example, a temperature inversion—the air mass is very stable. The result is often a thick, flat sheet of stratus clouds that may stretch from hundreds of miles offshore to the Cascades and the Coast Mountains and possibly even farther east.

## Clues to the Approach of a Warm Front

The general rules for gauging whether a warm front is approaching are:

- Look for approaching clouds, usually from the southwest, west, or northwest.
- Look for flat, sheetlike clouds (stratus).
- Look for thickening, lowering clouds.
- Look for surface winds from the east to southeast.
- Look for a gradual decrease in air pressure.
- Look for an increase in air temperature after warm-front passage.

Under these circumstances expect steady, widespread precipitation. (Mariners must remember these are general guidelines and, depending upon variations in local terrain, may not always hold true.)

## Warm-Front Precipitation

There are no sure field rules for determining whether a warm front will produce snow or rain. But as a rule of thumb, assume that precipitation will remain as snow (even though it may not stick to the ground) down to approximately 1,000 feet (300 m) below the freezing level. Current and forecast freezing levels can be obtained from National Oceanic and Atmospheric Administration (NOAA) weather radio, Environment Canada weather radio, and in some of the print and broadcast media. We'll further discuss sources of weather information in chapter 7.

It's important to remember that the freezing level reported is usually the free-air freezing level; that is, the altitude at which temperatures drop below freezing if uninfluenced by terrain effects such as trapped cold air in river valleys or fjords or heating of the ground by sunshine.

Let's say, for example, that Environment Canada weather radio reports the freezing level at 1,000 feet (300 m). You're slogging along up Princess Louisa Inlet under a sky of dense gray clouds that have recently moved in. In this case, the precipitation is likely to begin as snow. As mentioned, this is an approximate rule that can be affected by a variety of factors. A locally heavy shower, for example, can lower the snow level to as much as 1,500 to 2,000 feet (about 450–600 m) below the free-air freezing level, because moderate to heavy precipitation can drag cold air farther down from the base of the cloud. I recall a very unpleasant diving trip one blustery January day in Colvos Passage, to the west of Vashon Island in Puget Sound. As we submerged it was raining; when we surfaced, snowflakes the size of dimes were drifting down. Getting out of our wet suits in an unheated cabin was no fun!

Warm-front precipitation tends to be spread unevenly. University of Washington researchers have determined that precipitation tends to be clumped in cigar-shaped bands running parallel to a warm front. These bands generally range in width from 5 to 20 miles (8–32 km). Therefore, the intensity of the precipitation will vary with time. Even a slow-moving, wet warm front will be marked by occasional decreases in precipitation, if not outright breaks.

Rain and snow aren't the only forms of precipitation produced by warm fronts, especially in fjords and in river valleys trending east–west, such as the valleys of the Columbia and the Fraser rivers. If the air beneath is below freezing (usually when winds are out of the east) and the overlying air is sufficiently warm to produce rain, the warm front can produce either freezing rain or sleet (fig. 16). If the layer of cold air is shallow, freezing rain

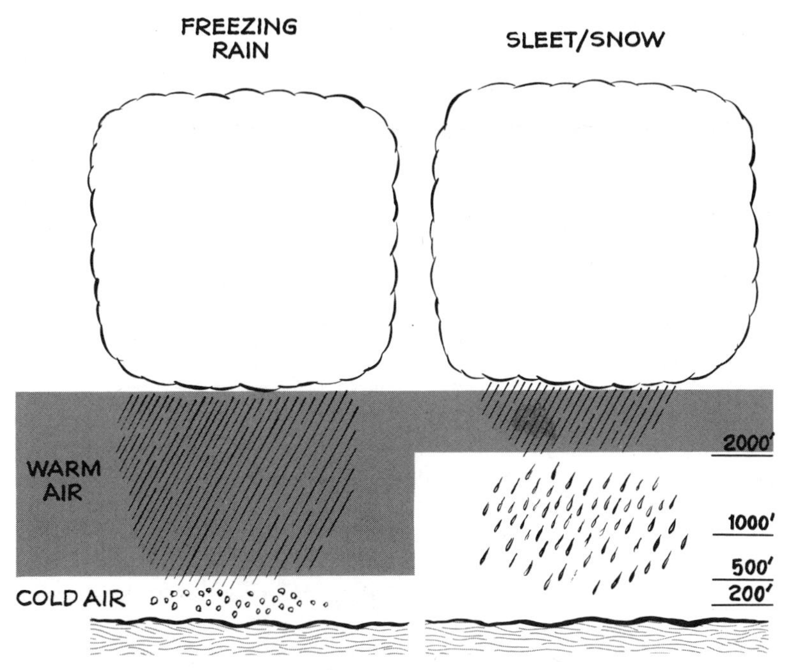


**Fig. 16.** Left, *freezing rain;* right, *sleet*

is most likely, whereas a thick cold-air layer is likely to produce sleet.

If, for example, you're tending your boat along the Columbia River with air temperatures below freezing, and the forecast freezing level is more than 2,000 feet (more than 600 m) above your elevation, then freezing rain is likely to glaze over everything exposed to the elements. However, if below-freezing temperatures are also reported at locations which are at higher elevations (places like Hood River Airport, for example), then the cold-air layer is thick enough that rain falling from the warm air aloft will probably freeze before it hits the ground. The result will be a good pelting with sleet or snow pellets.

## Cold-Front Clouds

Perhaps after you've spent a morning marooned in the marina lounge by precipitation from a passing warm front, the stratus clouds thin, possibly revealing streaks of blue sky. The air still feels relatively warm and moist. You decide to get under way, but just as you're rounding the breakwater, the skies darken more than ever, and rain or wet snow falls more intensely than before. The brief interruption in precipitation was simply a sign that the surface warm front had passed; the new precipitation signals the arrival of a cold front (fig. 14).

Inland areas east of the Cascades and the Coast Mountains often see more massive cloud buildups, a more definite break in precipitation, and some clearing (albeit hazy) between warm- and cold-front passage or, at times, only high cloudiness. But in the coastal regions, the transition from warm-front passage to cold-front arrival is usually brief, often marked by an accelerated drop in pressure and intensified precipitation.

In these transitions, warm, moist air arrives at the surface, so the air is no longer gliding up and over cooler air, and precipitation slows or stops. The air from the surface up is uniformly warm and moist. But then the cold air streaming down the back side of the approaching low collides with the warm, moist air already in place, abruptly thrusting it upward. If warm fronts are the turtles of the meteorological world, moving slowly but steadily, then cold fronts are the jackrabbits.

### *Cumulus Clouds*

The approaching cold air behind a cold front is much more dense, so the push upward can be very strong and fast, exceeding 20 miles per hour (32 km/h). The result is not layered clouds but a line of towering heap clouds, called *cumulus clouds.*

When such clouds produce precipitation or thunder and lightning, we call them *cumulonimbus clouds.* These can extend as much as 50,000 feet (roughly 15,000 m) above the surface of the earth. However, in the Pacific Northwest, where approaching cold air has usually been warmed by its passage over the Pacific Ocean, the temperature contrast between the approaching cold air and the warm air already in place isn't as great, and cumulonimbus clouds don't grow quite as tall. A maximum height of 20,000 to 30,000 feet (about 6,100–9,100 m) is more common in the Northwest, especially to the west of the Cascades and the Coast Mountains.

The exceptions to the rule of quick cold-front passage tend to occur when the cold front is sliding parallel to the coast or mountain ranges, with south or southwesterly winds in the upper atmosphere. This situation is most common near the Coast, Olympic, and Vancouver Island ranges and near the western slopes of the Cascade and Coast ranges.

## Clues to the Approach of a Cold Front

The general rules for gauging whether a cold front is approaching are:

- Look for clouds to thicken, lower, merge, and darken.
- Look for winds to increase, usually from the east or southeast, depending on terrain orientation.
- Look for a drop in pressure, usually rapid, then a rise after cold-front passage.

Under these circumstances, expect intensified precipitation with the front, colder temperatures after the front, and a wind shift to the southwest or west, depending upon terrain orientation.

## Cold-Front Precipitation

As the University of Washington research team had found along warm fronts, precipitation ahead of and along cold fronts tends to be organized in cigar-shaped bands parallel to the front. This precipitation tends to be more intense because of the rapid upward movement of the moist air ahead of and along the cold front (fig. 17).

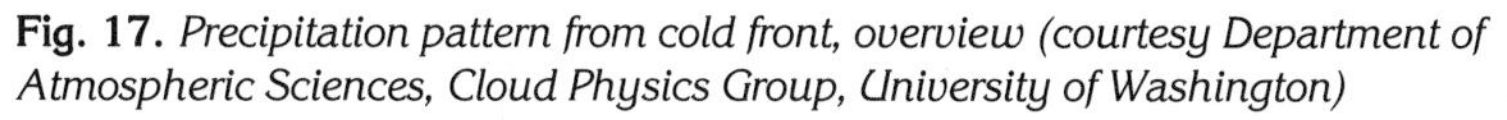
**Fig. 17.** *Precipitation pattern from cold front, overview (courtesy Department of Atmospheric Sciences, Cloud Physics Group, University of Washington)*

The upward movement is usually ten to a hundred times more rapid than that along a warm front. But whereas warm-front precipitation tends to be prolonged, often lasting as long as a day, cold-front precipitation is much briefer due to the front's more rapid movement. Precipitation associated with a cold front generally lasts only an hour or two.

We've already learned that the snow level usually extends 1,000 feet (about 300 m) below the freezing level when stratus clouds (warm-front clouds) produce the precipitation. But when the source is cumulus clouds (cold-front clouds), the snow level may extend as far as 2,000 feet (600 m) below the freezing level in heavy showers, especially after the cold front has passed.

Serious to severe thunderstorms are sometimes associated with the approach and arrival of cold fronts. This is true of the inland areas of the Northwest to the east of the Cascades and the Coast Mountains. Relatively warm, moist air is in sharp contrast to the cold, dry air moving down from Alaska, the Yukon, or the Northwest Territories. But in coastal regions, where the proximity of the Pacific Ocean moderates temperature contrasts, we tend to find thunderstorms after, not before or during, cold-front passage.

East of the Cascades and the Coast Mountains, clearing and drying usually follow the passage of a cold front. This isn't always the case for the areas to the west of these ranges. The areas that receive precipitation can be widespread or very localized, depending upon the delicate interplay of terrain features and wind.

## Occluded Fronts and Precipitation

An important variation on warm and cold fronts is the *occluded front* (fig. 18). Because of the sandwiching of warm and cold air, an occluded front combines the precipitation characteristics of both warm and cold fronts.

If you're experiencing prolonged rain or snowfall with occasional strong bursts, and possibly lightning and thunder too, an occluded front is the most likely culprit. Depending upon the direction of movement of the overall weather system, the passage of occluded fronts, like cold fronts, will be much more rapid than that of warm fronts. Because the occluded front is typically closer to the low than a cold or warm front, surface winds tend to be strongest—another good reason to avoid close encounters of the occluded kind.

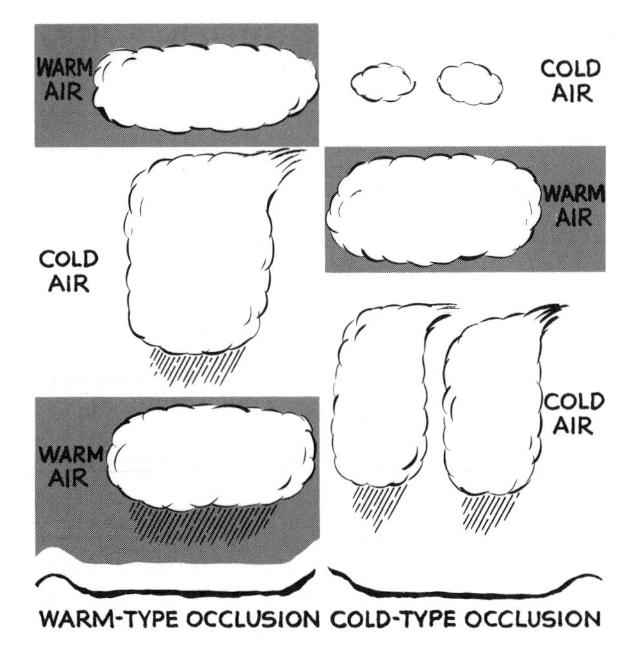


**Fig. 18.** *Precipitation pattern from occluded fronts:* left, *warm-type occlusion;* right, *cold-type occlusion*

CHAPTER 3

# The Interaction of Winds, Waves, and Currents

*The continual eddies and whirlpools, now favorable, now contrary, retarded one schooner and advanced the other, rendering it impossible to steer, and carrying us at their mercy.*

The log of Spanish explorer Galiano, aboard the *Sutil*, Sonora Island, British Columbia, 1792

Rarely does the skipper find "perfect wind." Either there's too much or too little, or when the wind is perfect for a sail, it's in the wrong direction for an easy approach to the slip afterward. If dealing with the wind doesn't provide enough challenge for our nautical skills, then its effect on the surface of the water will. Because water is a fluid medium, the movement of air over the water stirs it up into waves, which can range in size from gentle ripples to ship-swallowing monsters.

Wind isn't the only force that generates waves. There are four such forces: changes in air pressure, the tidal attraction of the sun and moon, seismic disturbances such as earthquakes, and wind. We'll largely focus our attention on waves generated by wind and the way such waves interact with currents. This knowledge provides a foundation for assessing *sea state,* using it to your advantage, and avoiding hazardous conditions. Sea state refers both to *sea swell* and to the *significant height* of the *wind waves.* The definitions of these terms are given later in this chapter.

## Wind-Generated Waves

The power of wind-generated waves is enormous. A single 4-foot-high wave expends 35,000 horsepower of energy per mile of coastline. That's one wave. A lighthouse along the Oregon coast once sustained roof damage from a 135-pound boulder flung upward by wave action; the roof was 91 feet above sea level! Such waves are the result of sustained high winds blowing across the water. Initially, as the wind starts to blow, the water begins to oscillate up and down. This is the birth of the wave.

The top of a wave is called the *crest* (fig. 19). The valley between two of the crests is called the *trough*, and the vertical distance between the

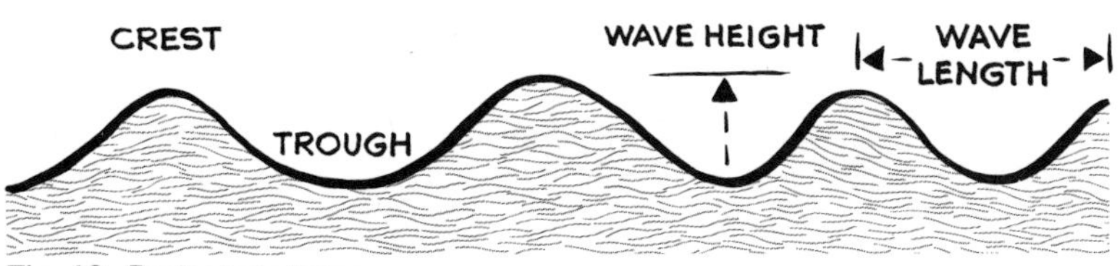


**Fig. 19.** *Basic wave terms*

trough and the crest is called the *wave height.* Another measurement of interest is the horizontal distance between the two crests, which is called the *wave length.* The time interval between the passage of two crests is called the *period.*

Oddly enough, the water particles or molecules don't actually travel with the wave that you see pass; they simply move in a circular orbit. It is the energy of the wave itself that moves. You can see this if you watch some flotsam (floating debris) bob in one place as waves move through. The exception is when the wave moves into shallow water, which causes the wave to break, or topple over.

## Estimating Waves and Winds Using the Beaufort Scale

We can describe and visualize how the surface of the water changes as the wind increases in strength, using terminology developed for the scale created in 1808 by Admiral Sir Francis Beaufort of the British Royal Navy (fig. 20). We can then use these visual cues to help judge wind speed, sea state, and the trend of the weather while ashore or under way.

The ability to recognize the changes that occur with increasing wind speeds is a useful tool for anyone who ventures very far from land. And for the crews of racing yachts or cruisers and of commercial vessels, that skill is a necessity. Needless to say, gale-force winds don't always wait to exist until we hear them reported on weather radio.

The Beaufort Wind Scale is calibrated in numbers that are known as forces, from force 0 to force 12—from calm seas to a mighty storm.

At the level of force 1, the mirrorlike surface of the water is creased by ripples that somewhat resemble the scales on a fish. Smoke drifts, but wind vanes remain still. Wind speeds generally range from 1 to 3 knots, called a *light air.* (One knot—that is, one nautical mile per hour—is equivalent to about 1.15 miles per hour or 1.8 kilometers per hour. Ten knots equals roughly 12 miles per hour or 18 kilometers per hour.)

As the wind continues to blow, reaching speeds of 4 to 6 knots, the ripples grow into small wavelets. You begin to see crests that form the peak or top of the wave, but they're still glassy in appearance. Leaves begin to rustle, wind vanes begin moving, and you feel the wind on your face. This is a force-2 wind, called a *light breeze.*

At 7 to 10 knots, the small wave crests occasionally break, creating *white horses.* The waves begin to broaden in length but still are not very wide. Leaves on trees are in constant motion; small flags are extended. This marks force-3 winds on the Beaufort scale, called a *gentle breeze.*

At 11 to 16 knots, the waves lengthen and now resemble little rows, and probably half of the wave crests are marked by white horses. On land, dust and loose paper swirl about, and small branches sway. This is force 4, called a *moderate breeze.*

Force-5 winds, a *fresh breeze,* range from 17 to 21 knots and are marked by longer waves, most with white horses, and perhaps a little spray occurs. Ashore, small trees sway, and small waves form even on ponds or slow-moving streams.

At 22 to 27 knots, waves continue to lengthen as well as increase in height, and spray flies from some of the crests. White horses or foamy crests are extensive. This is force 6, called a *strong breeze*—and it moves us into the range of small-craft advisories. On land, large branches sway, and power lines whistle.

As the wind speed increases to the range of 28 to 33 knots—force 7, a *moderate gale*—the sea becomes heavy with extensive spray. Some of the spray is blown into streaks paralleling the direction of the wind. Whole trees are in motion along the shore, and it becomes unpleasant to walk against the wind.

Force 8, a *fresh gale,* is the range from 34 to 40 knots, in which high waves begin to break well away from shore and the water surface is marked by extensive wind streaks. Twigs break off trees, and walking into the wind is difficult.

As the wind increases in speed to a range from 41 to 47 knots—force 9, a *strong gale*—the surface of the water is covered by dense foam streaks, and visibility is reduced. The whole sea surface begins to roll. Ashore, slight structural damage may occur to buildings.

Force 10 encompasses winds from 48 to 55 knots—well within the range of *storm warnings.* High waves now have overhanging crests, and visibility is markedly reduced. Trees may be blown over on land, and considerable structural damage is likely.

At 56 to 63 knots, a force-11 storm, the sea is covered with white spray. Small or medium-size vessels are occasionally lost to view by the wave action.

At force 12, a storm with winds of 64 knots or above, the sea and air are both white with spray, and visibility is essentially nil.

You can use the Beaufort scale to anticipate the likely sea state with any given wind forecast or report, or to judge wind speeds from the sea state that you observe. When I received my commercial seaplane pilot's license, I was required to memorize the scale's wind speeds and sea states and the corresponding visual clues on both water and land. That knowledge has come in handy more than once. With practice, you'll become skilled at estimating winds. Make a game out of it. Challenge your crew members to estimate the winds from the height and appearance of the waves or from the behavior of flags, smoke plumes, or trees on land.

**Fig. 20.** *The Beaufort Wind Scale*

| *Scale Number* | *Wind Speed* | *Common Name* | *Water Surface* | *Land Effects* | *Wave Heights** |
|---|---|---|---|---|---|
| 0 | <1 kt | Calm | Mirrorlike | None | Flat |
| 1 | 1–3 kt | Light air | Scalelike ripples | Smoke drifts | <1' |
| 2 | 4–6 kt | Light breeze | Small wavelets, glassy crests | Wind felt on face, leaves rustle | <1' |
| 3 | 7–10 kt | Gentle breeze | Few crests begin to break | Leaves in motion, small flags in motion | 2' |
| 4 | 11–16 kt | Moderate breeze | Wave rows lengthen, half of crests break into white horses | Dust swirls, small branches sway | 3' |
| 5 | 17–21 kt | Fresh breeze | Most waves marked by white horses, some spray | Small trees begin to sway | 4' |
| 6 | 22–27 kt | Strong breeze | Extensive white horses, spray common | Large branches sway, whistling heard in wires | 5' |
| 7 | 28–33 kt | Moderate gale | Extensive spray, streaks blown parallel to wind | Whole trees sway, walking against wind unpleasant | 6' |
| 8 | 34–40 kt | Fresh gale | Waves break away from shore, extensive streaks | Small branches break, walking against wind difficult | 8' |
| 9 | 41–47 kt | Strong gale | Whole surface rolls, covered by extensive foam | Some structural damage to buildings | 9' |
| 10 | 48–55 kt | Storm | Poor visibility, waves have overhanging crests | Large trees fall, structural damage to buildings | 10'+ |
| 11 | 56–63 kt | Storm | Sea white with spray | Structural damage | 10'+ |
| 12 | 64+ kt | Hurricane/ Storm | Sea white with spray, visibility nil | Structural damage | 10'+ |

*Wave heights are listed for areas where the fetch (the distance the wind can blow uninterrupted over the water) is 10 nautical miles or less, corresponding to most of the inland waterways of the Pacific Northwest. Waves will be higher in areas where the fetch is greater.

**Fig. 21.** *Definitions of official advisories/warnings*

| *Advisory/Warning Type* | *Wind Speed Range* |
|---|---|
| Small-craft advisory | 21 to 33 knots |
| Gale warning | 34 to 47 knots |
| Storm warning | 48 knots and above |
| Hurricane warning | 64 knots* |

*Not issued for the Pacific Northwest because hurricanes are warm-core storms, tropical in origin, that do not occur in the Northwest. Storm warnings are issued for Northwest disturbances with winds of 64 knots and above.

These are the common terms used in official warnings and advisories issued by the National Weather Service in the United States and by Environment Canada.

# Wave Heights

A whole list of factors go into determining how high a wave will be. You can't make hard-and-fast wave-height forecasts based on only one factor, such as the speed of the wind. Wave height is determined in part by the *wind speed* but also by its *duration;* that is, how long it has been blowing. *Fetch,* the distance the wind can blow uninterrupted in one direction over the water, is another important factor. The *width* of the body of water is another ingredient; the wider the body of water, all other things being equal, the bigger the waves will become.

Increasing *water depth* allows larger wave heights. And the colder the *air temperature,* the more dense the air, giving it greater impact on the water, and thus generating higher waves. *Rainfall,* especially heavy rainfall, tends to reduce wave height. *Rock reefs* cause waves to break offshore, just as an abrupt shoreline marked by cliffs generates much rougher water than a gradually sloping shoreline made of sand or mud. *Currents* also affect wave heights.

## Understanding and Using Wave Forecasts

When the term *sea state* is used in forecasts or warnings, it refers both to sea swell and to what is called the *significant height* of the wind waves. The significant height is simply the average height of the largest third of the wind waves observed or expected in an area. The wind waves are created by local winds; the swell, on the other hand, is composed of waves that were generated elsewhere and have traveled long distances, sometimes thousands of miles.

To help understand the idea of significant wave height, let's work through an example. Six waves were observed with heights of 3, 3, 5, 5, 7, and 7 feet. A third of the waves had heights of 3 feet, another third had heights of 5 feet, and the final and largest third of the wave had heights of 7 feet. In this simple example, 7 feet is the average height of the largest third of the wind waves observed in the area and, therefore, is the significant wave height.

Understanding significant wave height may seem to be of little interest to the cruising or racing sailor or skipper, and something best left to oceanographers. But mariners should know what significant wave height means and how it relates to the overall collection of waves found at sea, especially if they will be making passage over an exposed stretch of water.

**Fig. 22.** *Estimation of wave heights from significant wave height*

Significant Wave Height (SWH) = Average of Largest 1/3 of Waves

Most Frequent Wave Height = SWH x 0.5

Height of Highest 10% of Waves = SWH x 1.3

Highest Wave Height = SWH x 1.9

Significant wave heights can be used to calculate other handy wave heights (fig. 22). Multiplying the significant wave height by 0.5 yields the most frequent wave height you can expect to encounter. If the significant wave height is 6 feet, the most frequent wave height is 3 feet. The highest 10 percent of the waves is about 8 feet (determined by multiplying 6 times 1.3). And the highest wave would likely be about 11 feet (determined by multiplying 6 times 1.9).

Wave heights aren't always reported or available, but skippers can make an eyeball estimate of heights while under way—although this procedure isn't easy or exact. The best technique I've found is based on knowing the height of your eyes above the waterline from the helm of your vessel. This height can be measured while you're dockside. Look directly at some nearby object, such as a piling, while seated or standing at the helm. Have a crew member mark the spot that's level with your eyes, and then use a tape measure to determine the height of that spot above the waterline.

Out on the water, if the waves come level with your eyes, the estimate of wave height is an easy one: it's the same as the height that you measured back at the dock. If the waves reach halfway from the horizon to your eye level, then the wave height is roughly half that figure, and so on.

You can also do your own forecasting of significant wave heights by using tables of estimates (fig. 23), as long as you have a wind forecast and know your open-water fetch (distance from the nearest upwind shore or weather front to your position). The tables give significant-wave-height estimates in feet for a variety of fetches. All the estimates apply to open-ocean conditions where currents are less than 1 knot.

These tables are pretty easy to follow. First, calculate from your nautical charts your open-water fetch. Use that fetch to select the proper table. For the sake of this example, let's choose a fetch of 10 nautical miles. Now let's imagine the winds are forecast to blow at 40 knots for 12 hours. Drop down the vertical column of wind speeds to 40 knots. Then move horizontally to the right to the figure under the 12 hours column. The likely significant wave height for your particular set of conditions is 6.5 feet.

**Fig. 23.** *Significant wave height estimates in feet, open-ocean conditions*

**Wind Fetch: 5 Nautical Miles**

| Wind Speed (Knots) | Wind Duration in Hours | | | | | |
|---|---|---|---|---|---|---|
| | 1 Hour | 3 Hours | 6 Hours | 12 Hours | 24 Hours | 48 Hours |
| 20 | 1.7 | 2.2 | 2.2 | 2.2 | 2.2 | 2.2 |
| 30 | 3.2 | 3.5 | 3.5 | 3.5 | 3.5 | 3.5 |
| 40 | 4.9 | 4.9 | 4.9 | 4.9 | 4.9 | 4.9 |
| 50 | 6.2 | 6.2 | 6.2 | 6.2 | 6.2 | 6.2 |
| 60 | 7.8 | 7.8 | 7.8 | 7.8 | 7.8 | 7.8 |

**Wind Fetch: 10 Nautical Miles**

| Wind Speed (Knots) | Wind Duration in Hours | | | | | |
|---|---|---|---|---|---|---|
| | 1 Hour | 3 Hours | 6 Hours | 12 Hours | 24 Hours | 48 Hours |
| 20 | 1.7 | 2.8 | 2.8 | 2.8 | 2.8 | 2.8 |
| 30 | 3.2 | 3.5 | 3.5 | 3.5 | 3.5 | 3.5 |
| 40 | 4.9 | 6.5 | 6.5 | 6.5 | 6.5 | 6.5 |
| 50 | 6.7 | 8.2 | 8.2 | 8.2 | 8.2 | 8.2 |
| 60 | 8.8 | 10.0 | 10.0 | 10.0 | 10.0 | 10.0 |

**Wind Fetch: 20 Nautical Miles**

| Wind Speed (Knots) | Wind Duration in Hours | | | | | |
|---|---|---|---|---|---|---|
| | 1 Hour | 3 Hours | 6 Hours | 12 Hours | 24 Hours | 48 Hours |
| 20 | 1.7 | 3.2 | 3.7 | 3.7 | 3.7 | 3.7 |
| 30 | 3.2 | 5.9 | 6.0 | 6.0 | 6.0 | 6.0 |
| 40 | 4.9 | 8.6 | 8.6 | 8.6 | 8.6 | 8.6 |
| 50 | 6.7 | 11.0 | 11.0 | 11.0 | 11.0 | 11.0 |
| 60 | 8.8 | 14.0 | 14.0 | 14.0 | 14.0 | 14.0 |

**Wind Fetch: 50 Nautical Miles**

| Wind Speed (Knots) | Wind Duration in Hours | | | | | |
|---|---|---|---|---|---|---|
| | 1 Hour | 3 Hours | 6 Hours | 12 Hours | 24 Hours | 48 Hours |
| 20 | 1.7 | 3.2 | 4.4 | 5.1 | 5.1 | 5.1 |
| 30 | 3.2 | 5.9 | 8.1 | 8.8 | 8.8 | 8.8 |
| 40 | 4.9 | 8.9 | 12.0 | 12.0 | 12.0 | 12.0 |
| 50 | 6.7 | 12.0 | 16.0 | 16.0 | 16.0 | 16.0 |
| 60 | 8.8 | 16.0 | 20.0 | 20.0 | 20.0 | 20.0 |

Wind Fetch: 100 Nautical Miles

| Wind Speed (Knots) | Wind Duration in Hours | | | | | |
|---|---|---|---|---|---|---|
| | 1 Hour | 3 Hours | 6 Hours | 12 Hours | 24 Hours | 48 Hours |
| 20 | 1.7 | 3.2 | 4.4 | 6.0 | 6.4 | 6.4 |
| 30 | 3.2 | 5.9 | 8.1 | 11.0 | 11.0 | 11.0 |
| 40 | 4.9 | 8.9 | 12.0 | 12.0 | 12.0 | 12.0 |
| 50 | 6.7 | 12.0 | 18.0 | 22.0 | 22.0 | 22.0 |
| 60 | 8.8 | 16.0 | 23.0 | 27.0 | 27.0 | 27.0 |

Wind Fetch: 200 Nautical Miles

| Wind Speed (Knots) | Wind Duration in Hours | | | | | |
|---|---|---|---|---|---|---|
| | 1 Hour | 3 Hours | 6 Hours | 12 Hours | 24 Hours | 48 Hours |
| 20 | 1.7 | 3.2 | 4.4 | 6.0 | 7.5 | 7.8 |
| 30 | 3.2 | 5.9 | 8.1 | 11.0 | 14.0 | 14.0 |
| 40 | 4.9 | 8.9 | 13.0 | 17.0 | 20.0 | 20.0 |
| 50 | 6.7 | 12.0 | 18.0 | 25.0 | 27.0 | 27.0 |
| 60 | 8.8 | 16.0 | 23.0 | 33.0 | 35.0 | 35.0 |

Wind Fetch: 500 Nautical Miles

| Wind Speed (Knots) | Wind Duration in Hours | | | | | |
|---|---|---|---|---|---|---|
| | 1 Hour | 3 Hours | 6 Hours | 12 Hours | 24 Hours | 48 Hours |
| 20 | 1.7 | 3.2 | 4.4 | 6.0 | 7.5 | 9.5 |
| 30 | 3.2 | 5.9 | 8.1 | 11.0 | 15.0 | 18.0 |
| 40 | 4.9 | 8.9 | 13.0 | 17.0 | 24.0 | 28.0 |
| 50 | 6.7 | 12.0 | 18.0 | 25.0 | 34.0 | 38.0 |
| 60 | 8.8 | 16.0 | 23.0 | 33.0 | 45.0 | 48.0 |

Wind Fetch: 1,000 Nautical Miles

| Wind Speed (Knots) | Wind Duration in Hours | | | | | |
|---|---|---|---|---|---|---|
| | 1 Hour | 3 Hours | 6 Hours | 12 Hours | 24 Hours | 48 Hours |
| 20 | 1.7 | 3.2 | 4.4 | 6.0 | 7.5 | 10.0 |
| 30 | 3.2 | 5.9 | 8.1 | 11.0 | 15.0 | 18.0 |
| 40 | 4.9 | 8.9 | 13.0 | 17.0 | 24.0 | 30.0 |
| 50 | 6.7 | 12.0 | 18.0 | 25.0 | 34.0 | 45.0 |
| 60 | 8.8 | 16.0 | 23.0 | 33.0 | 45.0 | 60.0 |

When cruising near open ocean, it's important to remember that fetch is not necessarily the distance to the nearest shore in the direction of the wind. If a front lies between your position and the upwind shore, the surface wind direction will likely change, and the true fetch will probably extend from that front to your position. This is where weather facsimile charts come in handy. Surface weather charts allow you to measure the distance from your position upwind to where the wind no longer blows from the same direction; that is, at the front.

For example, lets say you're cruising along the west coast of Vancouver Island near Pachena Point, and winds are out of the southwest. This would seem to give you unlimited open-ocean fetch. According to the table for wind fetch of 1,000 nautical miles (fig. 23), if these southwesterlies had been blowing at 30 knots for 12 hours, you should expect a significant wave height of 11 feet. But if a weak cold front lies 50 nautical miles to the southwest, winds probably shift in direction at the front. Therefore, the effective fetch would be only 50 nautical miles, which would produce a likely significant wave height of between 8 and 9 feet.

## Rogue Waves

If you spend much time around sailors or skippers, you're likely to hear about rogue or killer waves. They're the object of legend and fear. Such rogue waves are unusually high and steep, and they can flip over even a freighter like a bathtub toy. The largest wave ever scientifically measured was 67 feet high, and that was in the North Atlantic. That doesn't mean bigger waves haven't existed; they simply haven't been measured (or the unfortunate scientist never lived to tell about it).

It isn't possible to forecast when or where such waves will occur. But before you sell your boat and take up radio-controlled boating on a city duck pond, remember that such waves are an open-ocean phenomenon. Certainly before setting out, however, it's wise to estimate the likely size of waves you'll encounter and to consider whether they would be beyond your capability or that of your vessel.

## Combined Seas

Weather forecasts from NOAA or Environment Canada usually give two height figures: one for wind waves (the significant wave height) and one for swell. Which one should be used for planning? What's needed is a composite of the two, called *combined seas.* (You can use the wave height from the significant-wave-height tables if you don't have a current forecast for wind waves.) To estimate the combined height of the seas you're actually likely to encounter, take the square of the significant wind-wave height and add it to the square of the swell height, and then take the square root of that sum (fig. 24).

**Fig. 24.** *Estimating combined sea height*

$$\text{(Significant Wind-Wave Height)}^2 + \text{(Swell Height)}^2$$

For example, if the significant wind-wave height is 6 feet and the swell height is 8 feet, we have 6 squared (which equals 36) plus 8 squared (which yields 64). The sum of the two squares is 100, and the square root of that sum is 10—giving us an estimated combined sea height of 10 feet.

We can use the same formula to develop best- and worst-case estimates of the combined sea height. Given the significant wave height of 6 feet and forecast swell of 8 feet, let's estimate the most frequent combined sea height and then the highest likely combined sea height.

To find the most frequent wave height, first multiply the significant wave height by 0.5 (as specified in fig. 22) to get the number 3. Then follow the formula for combined sea height (fig. 24): square the 3 (to get 9), and square the swell height of 8 (to get 64). The sum of the two squares is 73, and the square root of that sum is 8.5—giving us an estimate of 8.5 feet as the most frequent combined sea height.

For a worst-case estimate, multiply the significant wave height of 6 by 1.9 (as specified in fig. 22), which results in 11.4 feet. Again, follow the formula for combined sea height (fig. 24): square the 11.4 (to get 130), and square the 8 (to get 64). The sum of the two squares is 194, and the square root of that sum is 14—giving us our estimate of 14 feet as the highest combined seas we could expect to encounter as long as conditions remain unchanged. Again, this does not apply for a rogue wave, but such a wave is very, very rare.

## Currents and Wave Steepness

Wave height is certainly an important consideration for sailors and cruisers, and bears careful watching. But in some cases, *wave steepness* becomes even more critical. When waves interact with current, and that's frequent in the Pacific Northwest, the result can be surprisingly steep waves. Such steep waves can be very dangerous.

If the *set* of the current (its direction of movement) is in the same direction as the waves, the wave height will decrease. But if it's in the opposite direction of wave movement or within a few degrees of the opposite direction, the wave height will increase, at times dramatically (fig. 25). The *drift* of the current (its speed) will also have a dramatic effect

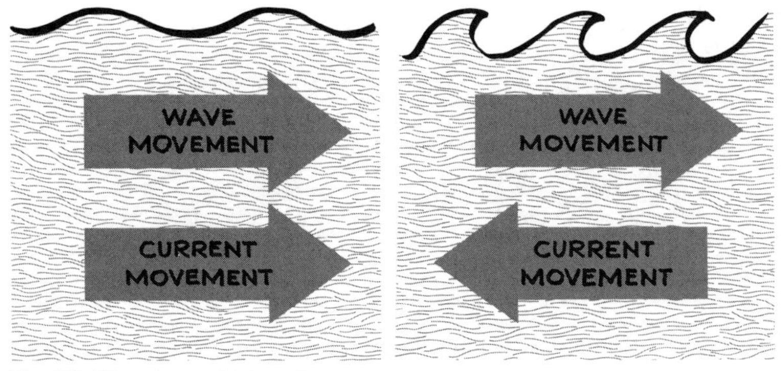


**Fig. 25.** *Wave/current interaction*

on wave height. If the drift of the current is in excess of 3 knots, expect wave height to double if the current is running in opposition to the wave direction. However, if the current drift is in the same direction as the movement of the waves, the wave height will probably be reduced by half.

At times, the combination of wind waves moving one direction and current moving another creates a churning motion called a *rip*. Since tidal currents are frequently strong, particularly in the San Juans and Gulf Islands, tidal rips are fairly common and resemble the spin cycle on a washer/dryer. The unfortunate skipper who becomes ensnared in such a rip may feel as though he or she is going for a ride in a Maytag, and vessels have occasionally capsized. Some small islands literally vibrate under the force of these tidal currents and rips.

Another twist to rips are the *overfalls* that are created when strong currents pass over submerged ridges. Such overfalls can produce an actual drop in the water level as well as severe turbulence. Imagine trying to run the Skykomish River in Washington (a favorite of whitewater fans) during spring snowmelt and you have an idea of what it's like to cruise through some of the rips and overfalls of the Northwest.

Wise skippers try to plan their passages for times when wind and current directions coincide. In areas where rips and overfalls are found, they make a special effort to move when current is at a minimum. Study your charts carefully, looking for areas of very changeable underwater terrain that coincide with speedy tidal currents. Such planning makes for a safer, speedier, and more pleasant cruise. It's also more fuel-efficient!

Also keep in mind the possible effect of your vessel's orientation in relation to the direction of the waves. Following seas—that is, waves approaching the stern—can make a vessel very difficult to control. Beam seas are waves approaching from either the starboard or port side of a vessel and in some situations can lead to capsizing. Quartering seas combine the worst of both beam and following seas (fig. 26).

**Fig. 26.** *Types of seas*

## Steep Waves at River Mouths

At the mouth of a river, the combination of an ebb tide and waves or swell can be dangerous, particularly near river bars. If the ebb-tide velocity is greater than 3 knots, expect the wave height to double, especially when the river discharge is high. Be especially cautious if the period or time interval between waves is 8 seconds or less, because hazardous conditions are very likely.

As an example, imagine that heavy rains have swollen the river discharge rate; tide and current guides forecast an ebb tide with a drift (velocity) of 4 knots just before the time of your planned departure. Timing the waves as they move past the breakwater, you determine the period (time interval) between wave crests is 7 seconds. That matches the criteria for hazardous bar conditions. Finally, remember that waves tend to look far less threatening from the back. So don't be caught by surprise if you're returning to harbor. Check the forecasts and reports just as carefully as before heading out, and consider delaying your return until closer to slack tide.

*Factors for Steep Waves at River Mouths*

- Ebb-tide velocity greater than 3 knots
- Wave period 8 seconds or less

## Other Conditions That Produce Steep Waves

Occasionally one group of waves collides with a second group of waves. This often happens when different currents meet or when swell from a distant weather system collides with local wind waves. It can also occur near a weather front. When waves collide, the resulting chop resembles a shark fin, steep and shaped like a pyramid. At best, such *cross waves* can make for uncomfortable going. At worst, they can be dangerous, making a small vessel uncontrollable. When waves reflect off a shoreline or breakwater and collide with incoming waves, the result is also much steeper and higher waves.

When waves roll into shallow water, a process called *shoaling* begins. The wave crests move closer to each other, eventually tumbling into surf. As waves begin to "feel bottom" and grow in size, they also turn toward the coast, becoming more aligned with the shape of the sea floor. This turning is called *refraction* and can change wave height as well as direction. That's because the incoming deep-water waves collide with the refracted waves that are turning to move into shore. This effect is most common along the coasts of Oregon, Washington, and British Columbia. Consult *The Coast Pilot* or *Sailing Directions* for your area for a discussion of hazardous shoaling or refraction effects.

CHAPTER 4

# The Wind: From Simple Strategies to Superstorm Survival

*A heavy wind began to blow from the southeast which by day break had increased to a violent storm. The master of the brig had all the sails taken in except the mainsail, under which we lay to. The tempest raged with unabated fury for three days. Our boat rolled with the swells from side to side in a terrifying manner, and at 1:30 in the afternoon, was already filled with water. With the arms in our hands we waited until a great billow came in, struck the boat and receded, then threw ourselves overboard and ran out on the shore beyond the reach of the water.*

Purser Timothy Tarakanof, the wreck of the
*St. Nicholas,* Destruction Island, 1808

The coast of the Pacific Northwest is a boneyard for ships. Some were wrecked by incautious skippers who ventured too close to the rugged cliffs and rocky reefs, but many were outmatched by the weather, caught by a howling tempest of wind and waves. It is perhaps one of the most unforgiving sections of coast along North America, just as it is one of the most beautiful.

There is little room for error. A day may begin with a cloudless sky, a light breeze, and gently undulating waves, but within hours can be transformed into a violent mix of raging, shocklike seas, driving rain, and furious blasts of wind. The coast is usually the first part of this region's waters to experience foul weather, given the prevailing movement of storms from west to east. The relative lack of weather observations from the open ocean makes exact (or even close) timing of storms difficult, if not impossible.

Despite the chameleonlike character of coastal weather, many skippers still plot a course there, lured by a sense of adventure, by the coasts' rugged beauty and mystique, and occasionally by necessity. Such voyages can be conducted safely if preceded by careful attention to forecasts and continued with a keen eye to key weather indicators.

Even skippers who have no intention of venturing beyond the more sheltered inland waters such as Puget Sound need to be familiar with open coastal weather: the big systems that first hit the coast often move inland. Skippers must exercise a considerable degree of skill and planning simply for gunkholing around popular destinations such as Puget Sound, the San Juans, the Gulf Islands, Princess Louisa Inlet, and Desolation Sound. True, sea states rarely become as wild as along the coast, and it's much easier to find sheltered harbor—but weather can be just as dangerous along the inland waterways.

The added rub on inland waters is that abrupt and frequent variations in terrain cause equally abrupt and frequent variations in the weather. Such variations are often beyond the scope of a single forecast or even the available weather observations, so it's up to the skipper to know what to look for and what action to take. To paraphrase one of the federal air regulations I drill into my flying students, "the captain in command is directly responsible for and the final authority as to the safe and efficient operation of his/her vessel."

# The Winds at Work

To understand the nature of storms and to anticipate their development and movement, we must take a closer look at the forces that strengthen and steer them.

## Winds in the Upper Atmosphere

Wind moves differently in the upper reaches of the atmosphere than it does at or near ground level. In the upper atmosphere (fig. 27), two forces are involved. The first is the *pressure gradient* force—simply the tendency for air to move from high to low pressure. Normally, the greater the difference in pressure over some distance, the bigger the pressure gradient, resulting in faster movement of air. The second force is the *Coriolis force,* which results from the rotation of the earth. The Coriolis force deflects the wind to the right of its original path.

These two forces tend to balance each other in the upper atmosphere. Therefore, the wind—rather than flowing from high to low pressure—tends to snake between the major high- and low-pressure systems, undulating from west to east in the Pacific Northwest.

When upper-atmosphere winds exceed 50 knots, meteorologists refer to them as the jet stream. Below that speed, they're generally called steering currents. In either case, these winds serve to steer the weather disturbances at the surface of the earth into storm tracks. Different storm tracks bring different precipitation patterns (as discussed in chapter 1 and reviewed in chapter 8, on field forecasting techniques).

Winds in the upper atmosphere strengthen when the contrast be-

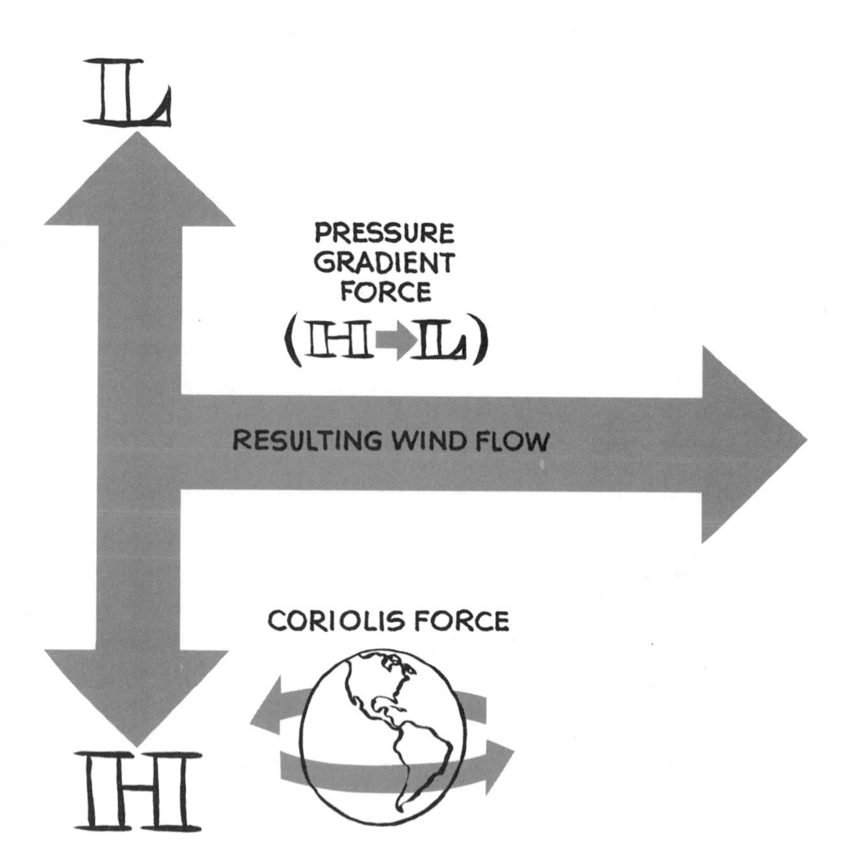


**Fig. 27.** *Wind flow in the upper atmosphere*

tween temperatures in the mid-latitudes (from 30 to 60 degrees) and those in the Arctic increases. That usually occurs between late autumn and early spring. The Arctic, thanks to minimal heating from the sun, plunges into the deep freeze. The stronger the jet stream or steering currents, the stronger the weather disturbances at the surface of the earth and the more likely they are to intensify.

Think of surface weather systems as riding on the steering currents of the atmosphere, somewhat like a boat on a river. For the purpose of this book, we can consider upper air winds to be those at and above 18,000 feet (about 5,500 m). Some television weather presentations show, at least in a general sense, the direction of the winds aloft.

One reason that weather maps of upper air patterns are so useful is that they indicate both where existing weather disturbances will tend to

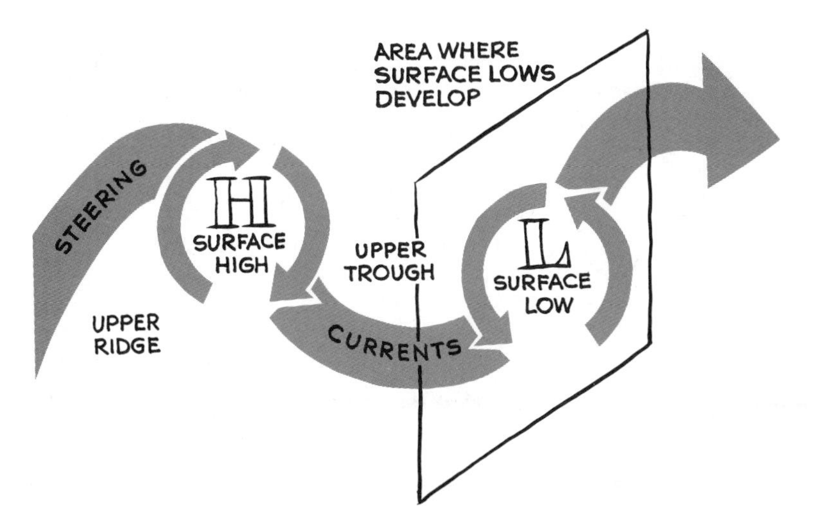


**Fig. 28.** *Relationship between surface systems and winds aloft*

move and where new pressure systems will likely develop. Surface high-pressure systems tend to be downwind of an upper ridge of high pressure, and surface low-pressure systems tend to be downwind of an upper trough of low pressure (fig. 28). Even if an upper trough hasn't produced a surface low-pressure system, one can develop in the area enclosed by the box in figure 28. Seeing such an upper trough just offshore on weather maps should lead to caution and restraint in venturing out from safe harbor.

Fast winds high up in the atmosphere tend to produce fast winds at the surface. Along the West Coast, watch the cirrus clouds that arrive in advance of an approaching weather disturbance. If you can see those clouds moving, the winds aloft are probably blowing in excess of 100 miles per hour (160 km/h). Expect strong, gusty winds soon.

## Surface Winds

The direction of surface winds is very different from the direction of winds aloft. Near the surface, the movement of air is somewhat more complicated than high up in the atmosphere. The pressure gradient and Coriolis forces are still present, but now friction comes into play.

The movement of air over water or land isn't smooth; the rougher the surface, the more turbulent the air flow over it, and the greater the friction. This upsets the balance between the pressure gradient and Coriolis forces. Air no longer flows parallel to high- and low-pressure systems. It now moves from high to low pressure, traveling clockwise out of and around a surface high and counterclockwise around and into a surface low. This tendency is reflected in a simple rule for locating high- and low-pressure systems: *If you stand with your back to the wind, low pressure will be to your left.* Figure 29 shows why this works.

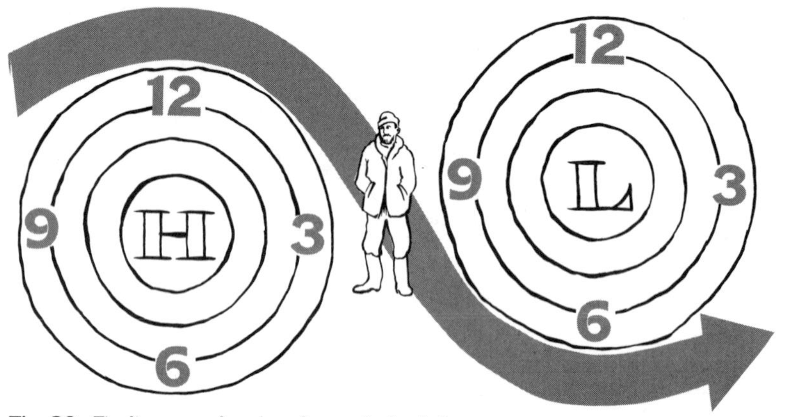


**Fig. 29.** *Finding a surface low from winds aloft*

Once the relative direction of a low is determined using this technique, your ship's compass can be used to find the magnetic direction. You know that steering currents move surface weather systems from west to east in the Pacific Northwest. Therefore, if the low is generally to the west of your position, it is likely moving toward you. If the low is to the east of you, it is usually departing.

This technique works well along the open coast and ocean, but it doesn't work quite as well in channels or inland waterways, such as Hood Canal in Washington or Trincomali Channel in British Columbia's Gulf Islands. That's because variations in wind direction in such areas are often caused by air being deflected or channeled in or around surrounding peaks, ridges, or cliffs. Use the technique only along more open waterways.

Just as the direction of winds aloft plays a major role in determining our weather, so does the direction of winds at the surface. Surface wind directions also provide valuable clues to the movement of storms. Because east to southeasterly surface winds are found ahead of a low, that tends to indicate a low may be approaching; because southwesterly winds are found behind a low, that tends to indicate a low is departing.

While a weather radio should be considered mandatory safety equipment aboard a boat, it has its limitations. The observations available for broadcast may not include a location close to your plotted course. That's why it's so important to learn the key indicators of approaching weather disturbances so you can analyze conditions at your location.

Principal Pacific Northwest wind indications include:

- Surface winds from the north or northeast: fair weather likely.
- Surface winds switching to the east or southeast: a weather disturbance is probably approaching. Expect southwest to west winds with the passage of the front.
- Surface winds switching from east or southeast to southwest or west: expect brief clearing, but be prepared for the possibility of showers or thundershowers.

The best and easiest way to keep track of significant wind changes is to use your compass. Note the direction of the wind relative to your vessel, and compare it to your compass heading. Make a mental note of that wind direction, or, even better, jot it down in a small notebook or on the margin of your nautical chart along with the time of your observation. I prefer to slide my charts into the plastic chart slickers available at most marine supply outlets and then mark down weather observations on the chart slicker with a grease pencil. That way the information is available for reference, can be washed off later, and doesn't mark up an expensive chart.

Changes in wind direction can signal changes in weather systems, but then again they may be due simply to variations in terrain. Look for other confirmation, such as changes in wind speed. An anemometer that precisely measures wind speed is nice, but not a necessity; just be alert to your environment. For example, is there an increase in the ripples or waves on the water, or are the boughs of trees along the shore rustling more forcefully?

### *Air Pressure*

Changes in wind direction and wind speed offer important clues to the approach of a major blow, but changes in air pressure probably offer the number-one clue to changing weather. That's because changes in pressure lead to changes in wind speed as well as direction. Careful attention to air pressure changes can provide early warning of impending high winds.

A ship's barometer is more than an attractive decoration for the cabin; it is an important safety device. If you charter boats, or own a small vessel such as a day sailor, ski boat, or kayak, consider purchasing one of the pocket barometer/altimeters available. There are digital models that will indicate not only current air pressure but also pressure tendencies over the past hours.

There are no absolutes as to which sustained pressure drops are significant, but the following guidelines have proven useful in my experience both in the forecast office and out on the water.

| *Pressure Change over a 3-hour Period* | *Recommended Action* |
|---|---|
| .02–.04" (.6–1.2 mb) | None. Continue to monitor sky, wind direction, and wind speed. |
| .04–.06" (1.2–1.8 mb) | Watch sky carefully for thickening, lowering clouds. Is the wind increasing, shifting to east or southeast? |
| .06–.08" (1.8–2.4 mb) | Watch sky and wind carefully, as above. Consider finding safe harbor due to the possibility of high winds. Conditions often merit small-craft advisories. |
| .08–.19" (2.4–5.9 mb) | Seek safe harbor immediately. Gale-force winds likely. |
| More than .2" (6.0 mb) | Seek safe harbor immediately. Storm-force winds likely. |

Checking your barometer every 3 hours will usually be enough to catch an approaching disturbance. When you experience 3-hour pressure falls that are equal to or greater than six-hundredths of an inch of mercury (1.8 mb), start checking the barometer every hour, and monitor weather radio continuously. At the least, small-craft advisory conditions are likely, with associated winds up to 33 knots. Such hourly monitoring will alert you to more rapid pressure falls that would precede either gale- or storm-force winds. The value of noting changes in barometric pressure can't be overstated. Fast-moving systems won't always give you a lengthy warning in the form of gradually lowering and thickening clouds, but there will almost always be a significant and sustained change in air pressure.

## Wind Phenomena

### *Coastal Winds*

As a low approaches the coast, winds begin to accelerate. The air essentially gets squeezed between the rugged terrain of the coast and the center of the low (fig. 30). Winds reported by ships in open ocean may be considerably less than those experienced by the pleasure and working vessels that are closer to the coast and thus subjected to this acceleration, called the *offshore wind maximum.* Expect winds to increase by as much as 50 percent over those reported when the low was farther away from land. The tricky part of this phenomenon is that these increased winds aren't picked up by coastal weather stations. The highest winds are found between 2 and 15 nautical miles offshore.

Meteorologists have also found that when an approaching low has winds equal to or greater than gale force (34 knots), the winds usually diminish shortly after the associated cold front moves through the coast. The important exception is when a high is building in behind the low. If the central pressure of that high is equal to or greater than 30.40 inches of mercury (1030 mb), strong winds will follow cold-front passage until the high arrives. Another situation also demands care: a deep low that approaches with winds *below* gale force. The structure of such a storm may have the strongest pressure gradient behind the cold front, saving the best (or worst!) for last.

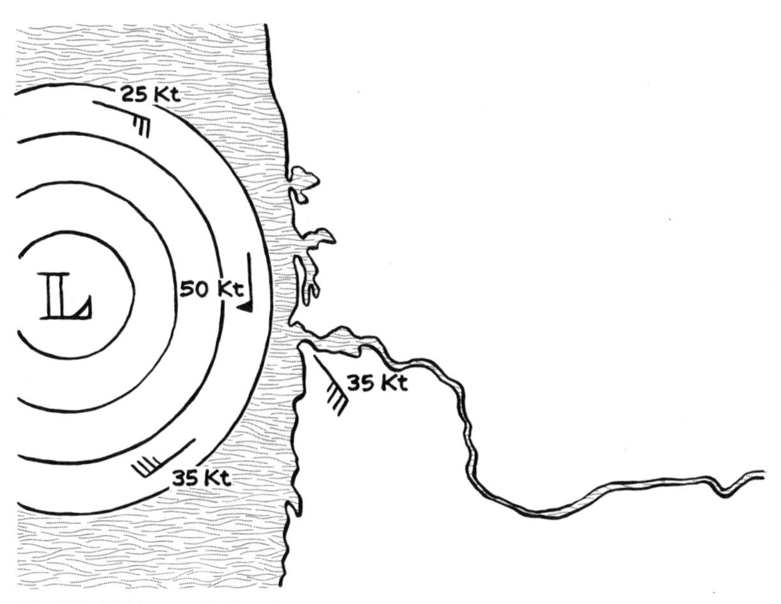


**Fig. 30.** *Offshore wind maximum*

## *Gap Winds*

Because of the rugged terrain that frames the waters of the Pacific Northwest, wind rarely blows uninterrupted. Often it's directed through gaps in the terrain, such as between two islands or two peaks—through a strait or a river valley.

When the wind is forced by terrain to move along the direction of a pass, strait, or waterway, meteorologists call that effect *channeling*. The same features that force the wind to change direction typically form a relatively constricted opening. As the air moves from open water into that gap, it tends to accelerate, often doubling in speed. The effect is the same as when you pinch a hose to increase the velocity of the water. This second

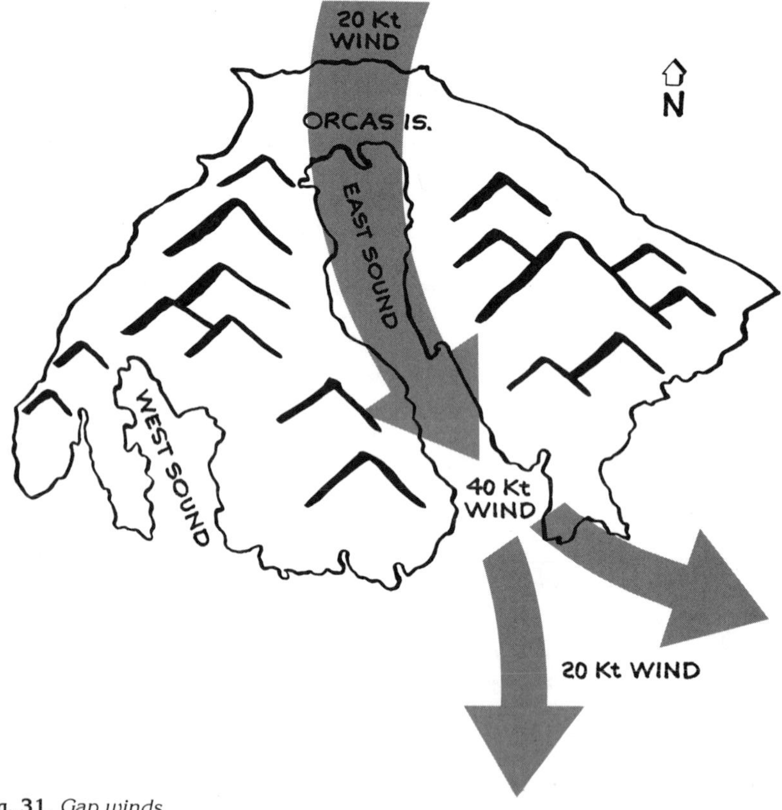


Fig. 31. *Gap winds*

effect is called *funneling*. Channeling and funneling work together to produce *gap winds*. The fastest winds, incidentally, are usually found near the downwind exit of the gap.

Orcas Island's East Sound presents a perfect example (fig. 31). The 20-knot winds we enjoy running before in the area between Orcas and Lummi islands can increase to 40 knots in East Sound! High terrain surrounds East Sound. Northerly winds accelerate through the gap, eventually slowing again near Blakely and Lopez islands. Southerly winds produce the same effect, but the acceleration is in the opposite direction, leading to rough conditions near Parker Reef just to the north.

Another obvious example of channeled winds is through the Strait of Juan de Fuca. Winds near the west entrance have reached 65 knots with gusts up to 90. Such channels or gaps also increase tidal currents. Gap winds directed against strengthened tidal currents produce steep, breaking waves, requiring extreme caution. At times the best decision is simply to wait until slack water, or at least until the wind and current are moving in the same direction.

### *Wind Blocking*

Another effect of terrain is to provide blocking or sheltering from the wind. As you pass close to the leeward shore of an island, wind is substantially reduced. But once you pass the island, a sudden increase in wind can produce a rude awakening. Keep your eye on the water ahead for sudden roughness; that's a good sign of changing wind conditions.

Winds moving around an island curve, joining up some distance from the leeward shore. In the case of a round island with the summit in the center, expect the air flow to rejoin at a distance from the island center approximately equal to five times the height of the summit (fig. 32). For example, just to the north of Texada Island in the Strait of Georgia is Harwood Island, rising to approximately 480 feet. Given westerly winds, expect sheltering up to a maximum of 2,400 feet (a little less than half a mile) offshore to the east.

To visualize the effect terrain has on wind, slide your chart into a waterproof cover and then use a grease pencil or a water-soluble marker to draw arrows depicting wind directions reported on weather radio. That will help you to see where gaps tend to accelerate wind, how wind tends to follow the shape of the land, and where you can expect to find some sheltering.

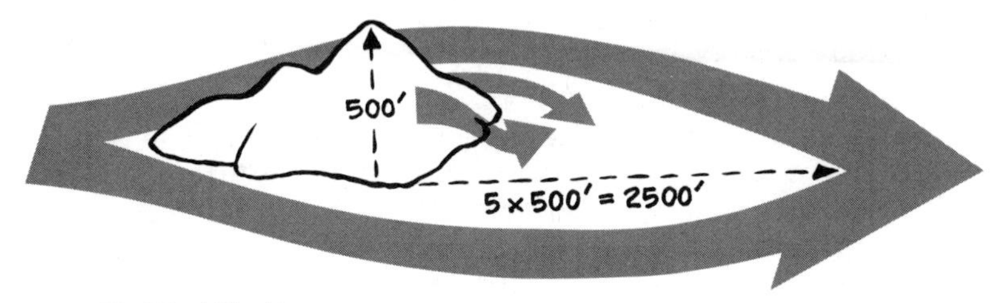


**Fig. 32.** *Wind blocking*

### *Corner Winds*

Air moving over land can also cause changes in wind speeds on either side of an island or peninsula, a phenomenon called *corner winds*. As wind speeds decrease over land, the winds also tend to turn toward low pressure. A southwesterly wind over water will change to a southerly over land. If you are sailing in an area where land is to your right and the wind is at your back, you can expect the change in wind direction over land to cause air to converge in your area, generating winds 20 to 25 percent

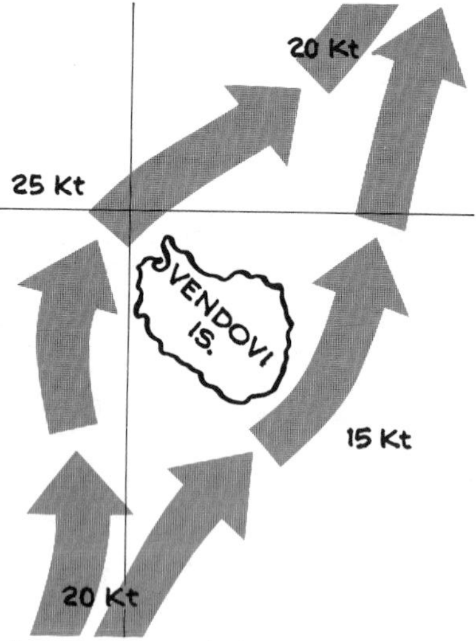

Fig. 33. *Corner winds*

stronger than on land (fig. 33). There's also a tendency to produce more clouds in this area because the converging air rises. Racing sailors tend to seek out this zone of higher wind speeds.

Cruisers or kayakers may wish to seek out the other side of the island or peninsula, where land is to the left with the wind still at your back. The same turning phenomenon caused by friction creates a split, or divergence, in the surface winds, making them lighter on the right side of the island (fig. 33). And the sinking air may reduce cloud cover, offering a better chance of sunshine!

### *Cliff Winds*

Steep shorelines can do interesting things to the wind and the water. Winds blowing over water against a steep cliff create a small zone where the wind blows in the opposite direction; that is, offshore. The gusty, confused winds also create a zone of confused, choppy seas (fig. 34).

Winds blowing over land toward water also generate a zone of gusty winds, suddenly blowing in the opposite direction, usually within a nautical mile or two of the cliff. That's followed by a band of lighter winds and calm seas, often known as *wind shadow,* which may extend another nautical mile or two toward open water. These zones of strong, gusty winds and light winds tend to remain in the same place as long as the wind direction and speed remain fairly constant. The size of these zones depends upon the strength of the wind and the steepness of the shoreline.

A knowledge of the effects of cliffs on wind can be very helpful to mariners seeking to avoid either choppy seas or calm winds. Comparing wind speeds and direction with terrain is the key. If nautical charts don't provide enough topographic information, pick up some topographic relief maps at an outdoor supply store. An evening spent reviewing such a map marked with forecast winds and the expected race course will be helpful to the racing sailor; it could mean the difference between just finishing and winning!

### *Land and Sea Breezes*

Most of the wind effects discussed in this chapter are typically found within a low or near the boundary of a high. That's where the pressure

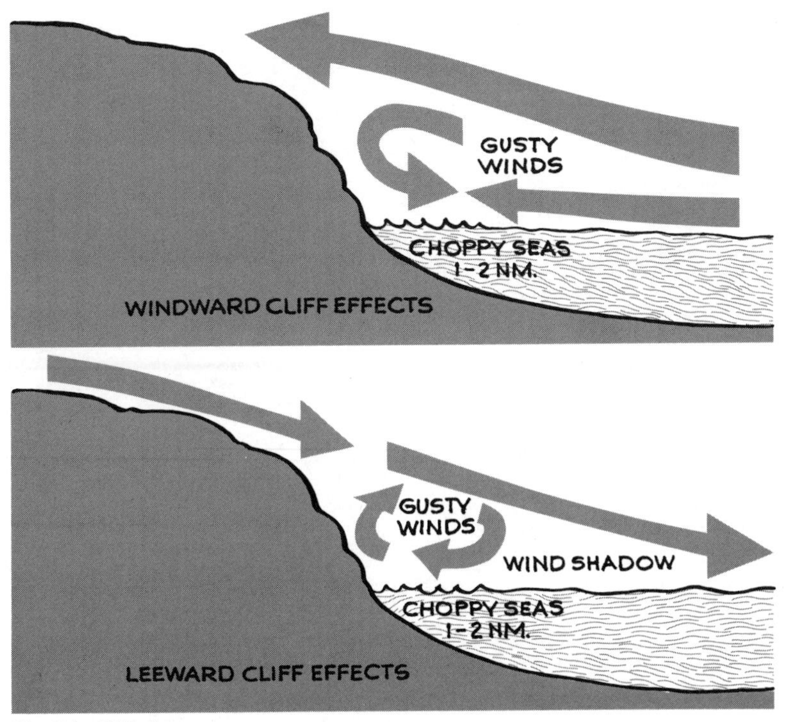


**Fig. 34.** *Cliff effects*

gradient, or difference in pressure over distance, is usually greatest. But significant winds can be found within a high, where the difference in pressure over distance tends to be small. An example is the sea breeze/ land breeze.

Let's say you own a sailboat moored at Shilshole Bay Marina in Seattle. Weather forecasts indicate high pressure dominates the area. Hoping for enough wind for a pleasant sunny afternoon sail, you look at the water in the morning and can't find even a ripple. It looks like glass. Should you cancel your plans because of the lack of wind? Not necessarily, thanks to the possibility of sea breezes.

This phenomenon is caused by the differing rates at which land and water absorb and radiate heat. During the day, land heats up much more than water, as you know from walking barefoot across pavement or sand on a hot day. The heat rises over the land, drawing in cooler air from over water. Because the air is moving from water toward land, meteorologists refer to this as a sea breeze (fig. 35). It can occur along lakes and rivers as well.

By mid-afternoon on a hot summer day, sea breezes can generate winds of 5 knots or more on a small body of water such as Oregon's Lake

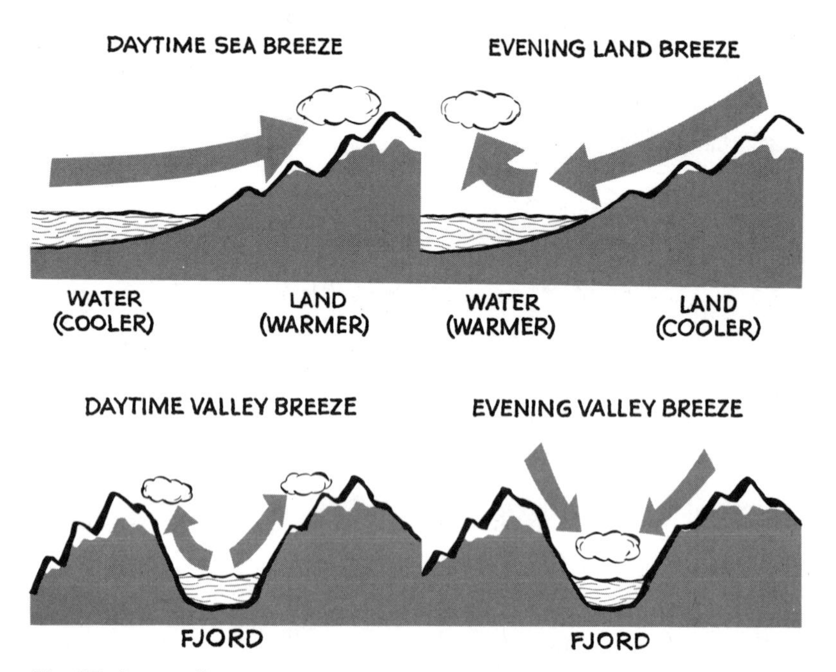


**Fig. 35.** *Breeze effects:* top, *sea breeze;* bottom, *valley (fjord) breeze*

Oswego and 10 knots or more on a large body such as Puget Sound. Along the coast, sea breezes can reach 20 knots or more. When sea breezes funnel through a gap or channel, wind velocities can easily double to afternoon peaks of 30 or 40 knots. Such sea breezes normally begin about 3 to 5 hours after sunrise, reach a maximum during the afternoon hours, and taper off right after sunset. Sea-breeze direction tends to be northerly over Puget Sound, westerly in the Strait of Juan de Fuca and the Columbia River Gorge, and northwesterly in the Strait of Georgia. Expect local variation due to terrain differences.

In the evening, land cools far faster than the adjoining water. The cool air over land now sinks and slides toward the water, generating a land breeze (fig. 35). Such land breezes are usually weaker than the daytime sea breezes. If buildings or higher terrain exist upwind, the wind will be fickle, changing direction as the air swirls through and over the gaps. Land breezes tend to reach a maximum during the late night hours.

Sailors can take advantage of these breezes by staying closer to shore and keeping an eye on the water to see where the roughness generated by the breezes ends. Plan to tack or jibe before reaching that boundary. Large paved areas ashore, such as parking lots or a major downtown area, help generate the strongest sea breezes, making them an ideal stretch of shoreline to choose for an afternoon sail. Suburban or rural areas ashore generate somewhat weaker sea breezes.

Breezes helpful to the sailor may be troublesome to the skipper of a small powerboat or to a paddler in a canoe or kayak, who would prefer calm conditions. In these cases, you can plan your trips for the early- to mid-morning hours, taking your break during the time of maximum sea-breeze strength in the early- to mid-afternoon hours. You can also head farther from shore, where the strength of sea breezes and land breezes diminishes.

### *Valley/Fjord Breezes*

In the Pacific Northwest, land breezes and sea breezes also can be generated on inland bodies of water. These are called valley (or fjord) breezes. During the day, the land heats and the air warmed by that heating rises up either side of a valley (whether or not that valley contains a body of water), spilling over the adjoining hills and ridge tops. This movement can amplify sea breezes on bodies of water surrounded by higher terrain, which describes just about every inland body of water in the Pacific Northwest. At night, the land cools. The cool air flows downslope in what is called a gravity wind, converging in the middle of the channel (fig. 35).

Valley breezes can be generated even during low cloud cover. The key is that the cloud cover must be thin enough to expose the upper slopes of the surrounding mountains to the sun. If the clouds blanket the surrounding mountains all the way to the ridge tops, the land won't have the opportunity to heat up any more than the water; therefore, no valley breeze will be generated. Another factor that prevents valley breezes from forming is the presence of winds of 10 knots or more during the morning hours. Such winds prevent the sea-breeze/land-breeze circulation from developing.

### *Breezes and Precipitation*

When moist ocean air dominates the Pacific Northwest after a cold front moves through, both time of day and location determine the likelihood of precipitation. Land, sea, and valley breezes are responsible for this pattern.

**Afternoon/Early-Evening Hours:** Because sea breezes flowing from water to land are most prevalent during the afternoon and early-evening hours, they tend to thicken clouds pushed up against surrounding mountains or hills (fig. 35). If the air is moist and unstable, this forced flow upslope can produce showers or even thundershowers. That's why the afternoon hours are the most likely time for showers or thundershowers along the slopes of the Cascades, the Olympics, and the Coast Range. The same is true for the slopes of British Columbia's Coast Mountains and the Vancouver Island Range.

**Later Evening/Night Hours:** After sunset, the flow of air is downslope (fig. 35), toward the Willamette Valley in Oregon, Puget Sound in Washington, or the Strait of Georgia in British Columbia. This reduces the chance of precipitation along the mountains and increases the chance of precipitation on or near the water. The showers that form up against the mountain slopes may actually drift seaward.

### *Lee-side Troughing*

Occasionally, much higher winds are found to the lee of mountains than can be explained by pressure gradients or even funneling effects. This was the case when the Hood Canal Bridge sank in Washington State in February 1979.

When surface winds blow out of the south to southwest, the Olympic Mountains may generate a small low-pressure system in the lee of mountains over Puget Sound, Hood Canal, Admiralty Inlet, or the Strait of Juan de Fuca. Some of the air striking the Olympics swirls around the sides, but some also flows over the top. As that air flows down the lee side of the Olympics, it hits the air flowing *around* the Olympics and actually rebounds slightly, helping create this lee-side low or trough (fig. 36).

As air flows through the channels and inlets of this area, it accelerates more than usual, generating higher wind speeds. I've found lowered clouds and intensified precipitation in these areas while flying both seaplanes and land planes. If coastal winds are from the south to southwest, consider avoiding this zone or, at least, proceeding with extreme caution. Although this effect has been documented only to the lee of the Olympics, it's entirely possible that strong winds may produce the same effect to the lee of other localized mountain ranges in the Pacific Northwest.

## Superstorms

A superstorm usually begins as a fairly benign low, far away in the tropics or subtropics. But the long journey to the Pacific Northwest can profoundly transform this low into a violent storm that threatens life and property.

There are lows that produce rain and wind and blustery weather—and then there are lows that could put some hurricanes to shame. By the strict definition of hurricane, we're spared that experience here in the Northwest. But that offers scant comfort. The superstorms we do experience produce gusts—and sometimes sustained winds—that exceed hurricane force. The Columbus Day storm of 1962, which hammered both western Oregon and Washington, was one example. The Inauguration Day storm of 1993, which swung from the mouth of the Columbia River up into Puget Sound, was another.

In a Northwest superstorm, barometer needles take a nosedive. An almost eerie calm changes into a low moan as winds begin to rise, whipping halyards into a singing frenzy at marinas and churning the water just beyond breakwaters into a frothy furor. Gusts can easily exceed 100

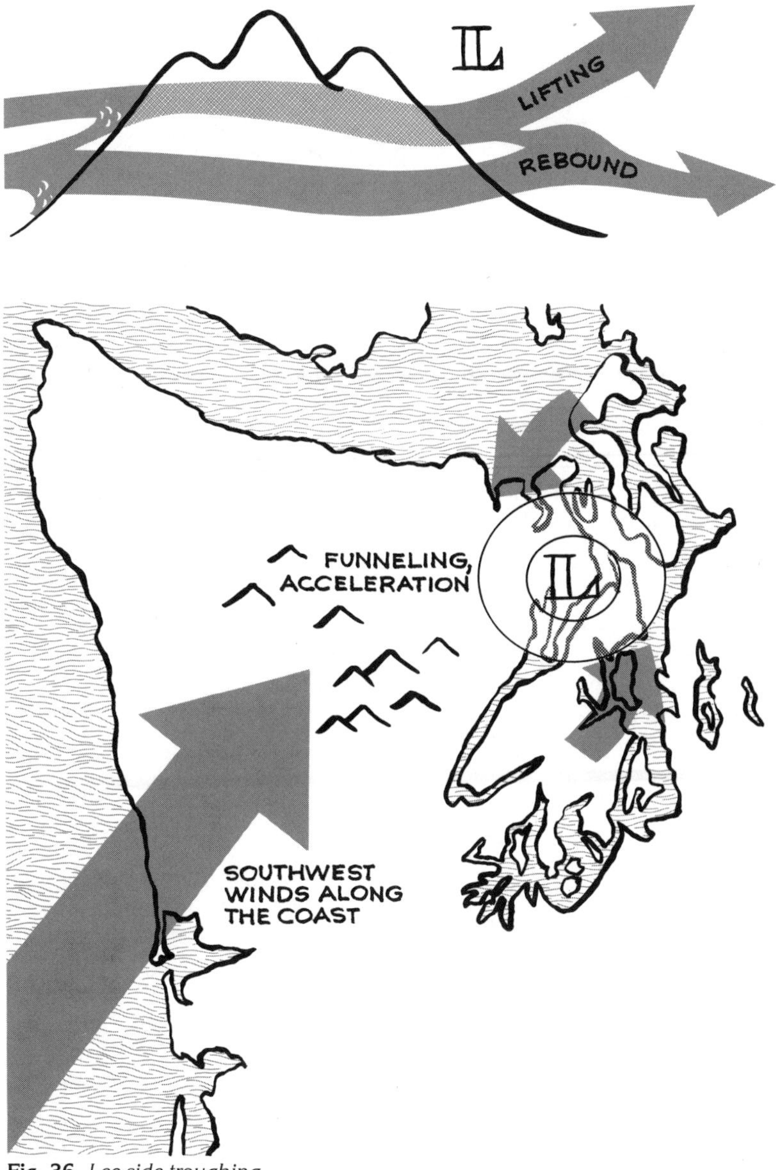


**Fig. 36.** *Lee-side troughing*

miles per hour at exposed coastal locations, such as at Solander Island along the west coast of Vancouver Island, at Cape Disappointment along the Columbia River, or at aptly named Destruction Island along the Washington coast. Inland areas aren't spared, as both the Columbus Day

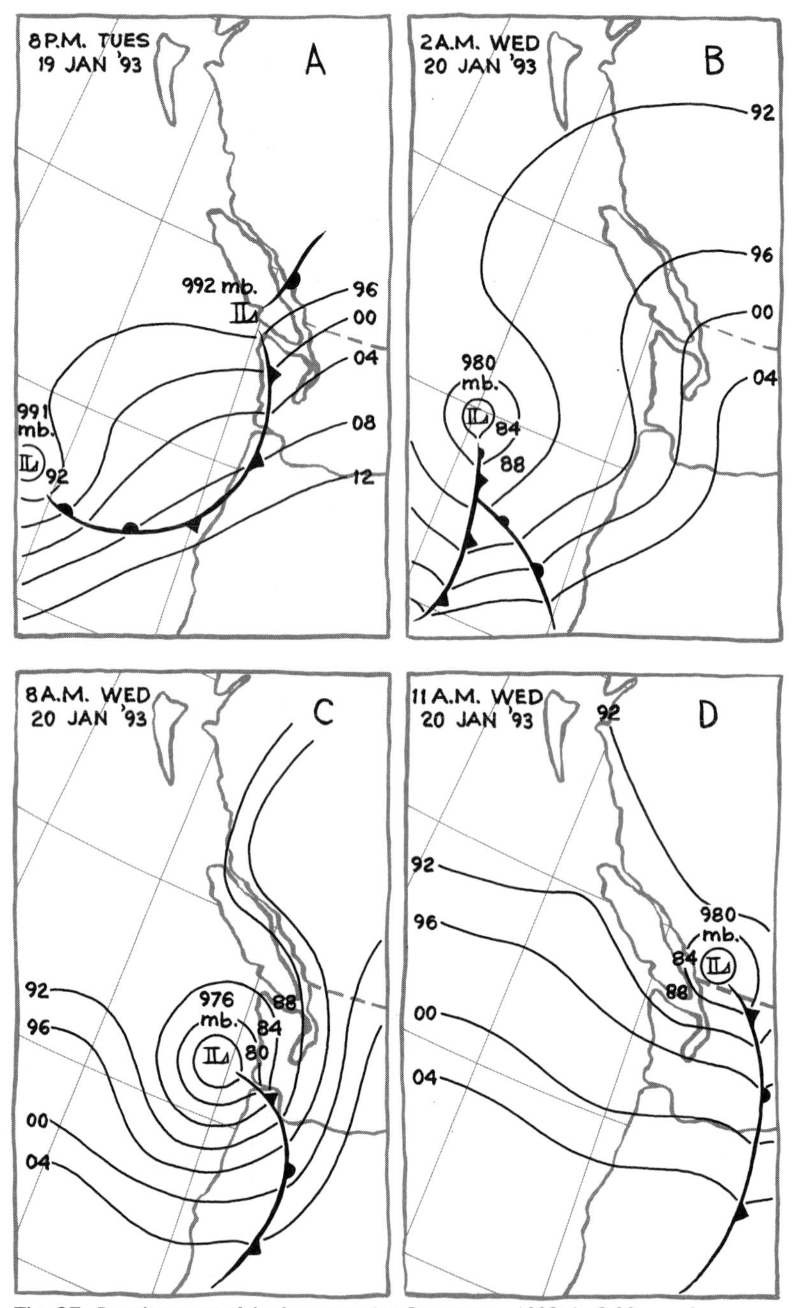


**Fig. 37.** *Development of the Inauguration Day storm, 1993:* A, *8:00* P.M., *January 19;* B, *2:00* A.M., *January 20;* C, *8:00* A.M. *January 20;* D, *11:00* A.M., *January 20*

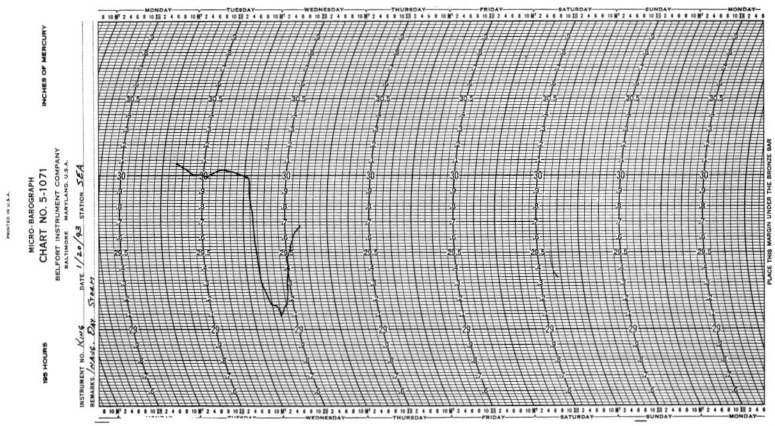

**Fig. 38.** *Barometric pressure record of the Inauguration Day storm, 1993*

and Inauguration Day storms produced gusts exceeding 80 miles per hour in Puget Sound. Such superstorms are still under extensive study and defy easy categorization or forecasting. But a couple of factors have been seen repeatedly and offer useful clues to skipper and meteorologist alike.

First, such storms tend to form between the months of October and February. The temperature difference then between the tropical and polar air masses is greatest, which in turn generates the most intense surface lows and the fastest jet streams. The two most violent storms of recent history in the Northwest both occurred during this period. The Columbus Day storm of 1962 was in October and the Inauguration Day storm of 1993 was in January.

Second, superstorms typically originate in the tropics or subtropics. The combination of the warm air and water temperatures in those southern latitudes supply a low with far more moisture and energy than could be drawn from the air and water farther north. When the jet stream dips below 40 degrees north latitude, the potential for one of these soggy lows to cross the jet, collide with colder air, and intensify into a monster is great. Once a subtropical low crosses the path of the jet stream, it can intensify rapidly, sometimes giving skippers less than 12 hours' warning of its transformation into a fire-breathing superstorm.

Notice the sequential surface weather charts showing the location and intensity of the low responsible for the 1993 Inauguration Day storm (fig. 37). Once it has migrated north from the tropics, it intensifies rapidly. The warning to skippers and landlubbers alike was clear on the barograph record of air pressure (fig. 38); a big blow was about to hit.

Keep a careful eye on television weather reports if you're ashore, or monitor NOAA or Environment Canada's weather radio broadcasts. Such storms evolve constantly, requiring frequently updated forecasts of movement and intensity. People who fail to keep up to date have only themselves to blame for damage or injury caused by such a lack of vigilance.

Whenever weather radio broadcasts warn of an approaching low with central pressure equal to or less than 28.80 inches of mercury (975 mb), pay close attention to all updates. If your ship's barometer indicates a 3-hour pressure fall matching or exceeding two-tenths of an inch of mercury (6 mb), be prepared to seek immediate refuge, and secure your vessel carefully.

Finally, be aware that although the lows moving out of the tropics or subtropics are the ones that produce superstorms in the Northwest, lows that develop in the Gulf of Alaska or the Aleutians can also produce hazardous weather in this region. They likewise deserve careful attention.

Chapter 5

# Special Hazards: Fog, Localized Precipitation, and Thunderstorms

*Fog from Seaward, Weather Fair;*
*Fog from Land brings Rainy Air.*

Sailors' proverb

## Fog

Fog—essentially a cloud at ground or water level—can cause any number of problems for a skipper. It can make accurate navigation difficult if not impossible, to say nothing of trying to avoid deadheads or other vessels. (A chance encounter between a 300-foot freighter and a 30-foot sailboat will have but one winner, and it won't be the sailboat!)

Five types of fog are most prevalent throughout the coastal areas of the northwestern United States and British Columbia: radiation, advection, warm frontal, sea, and steam. Before exploring the subtleties of each, it makes sense to understand the fundamentals common to all.

Recall the example of seeing your breath on a cold day. To review: as we inhale air, it warms close to our body's normal temperature of 98.6 degrees Fahrenheit (37° C) and is moistened. As we exhale, that warm, moist air is rapidly cooled to the temperature of the air around us and may condense from water vapor (a gas) to water droplets (a liquid). This is similar to the way steam from the shower condenses on a cold bathroom mirror. A similar process produces fog.

### Radiation Fog

Radiation fog is most common during the autumn months and often follows wet weather. Perhaps rain fell during your drive out to the marina, followed by nighttime clearing. The nighttime sky was dazzling, filled with stars. The next morning, you're surprised by dense fog and possibly drizzle when you go out on deck to check the lines. But after a cup of coffee or two, you notice the gray fog gradually thinning, the sun first appearing as a light disk, then turning a dazzling yellow.

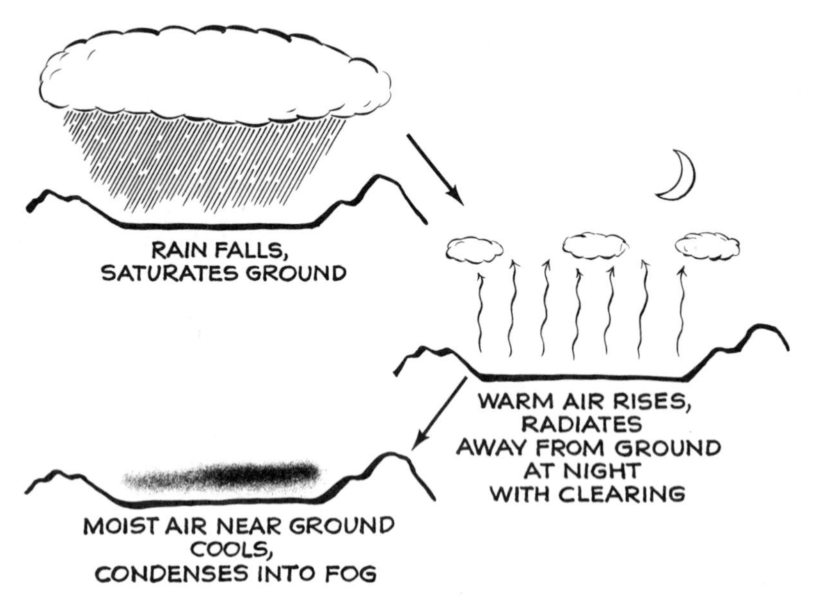
RAIN FALLS,
SATURATES GROUND
WARM AIR RISES,
RADIATES
AWAY FROM GROUND
AT NIGHT
WITH CLEARING
MOIST AIR NEAR GROUND
COOLS,
CONDENSES INTO FOG

**Fig. 39.** *Formation of radiation fog*

Radiation fog is produced when some of the moisture from damp ground evaporates into the lower layer of the atmosphere (fig. 39). As the disturbance that produced the moisture passes, the clouds clear. Overnight, considerable heat is radiated by the ground to the colder atmosphere above. The moist air close to the wet ground or water cools and condenses into a layer of fog.

Radiation fog is especially common in river valleys, fjords, and basins (such as the Puget Sound basin) because cold air tends to drain downslope, hastening the cooling and condensation of the moist air below. It is occasionally rather thin, often only several hundred feet deep. Even when radiation fog is thick, it often burns off by mid-afternoon, then progressively earlier each day thereafter. However, if the moist air is trapped close to the ground or water, the evaporation of the fog is much slower. This usually happens when warm air lies on top of colder air near the surface. Meteorologists call this situation an *inversion,* because air temperature normally cools with increasing altitude.

## *Radiation Fog Clues*

- Moist ground from rain or melting snow, or relatively warm water temperatures.
- Clearing that allows extensive overnight cooling, especially in the late autumn to early spring, when nights are long.
- Light winds, generally 5 knots or less.

## Advection Fog: The Onshore Push

While radiation fog is most common during the months when nights are long, advection fog (fig. 40) is most frequent during the summer months, primarily west of the Coast Mountains and the Cascades in British Columbia, Washington, and Oregon. The movement of advection fog into the interior is often referred to as an *onshore push*.

During the summer, daytime air temperatures are considerably higher than those of the ocean water just off the coast, often by 30 degrees Fahrenheit or more. The warm air rises, and cooler, moist ocean air surges in to replace it. That movement inland can be boisterous, with a sudden wind shift and wind speeds reaching gale-force strength. Once the moist air has moved inland, and the air temperature drops overnight, the ocean air condenses into fog or a low layer of flat stratus clouds. Advection refers to this horizontal movement of air from water to land.

Advection fog typically develops shortly before sunrise, surrounding the Coast, Olympic, and Vancouver Island ranges and blanketing the western slopes of the Cascades and the Coast Mountains to a maximum of roughly 5,000 to 6,000 feet (about 1,500–1,800 m) above sea level, but often much lower. Such stratus or fog rarely crosses the Cascades of Washington and Oregon or the Coast Mountains of British Columbia, so sun-seekers often trailer their boats (if possibile ) to destinations along the eastern slopes of these ranges, such as Washington's Lake Chelan.

Advection fog, like other weather phenomena, never moves in without warning, although the clues can be subtle. Following are the ones to watch for, in chronological order.

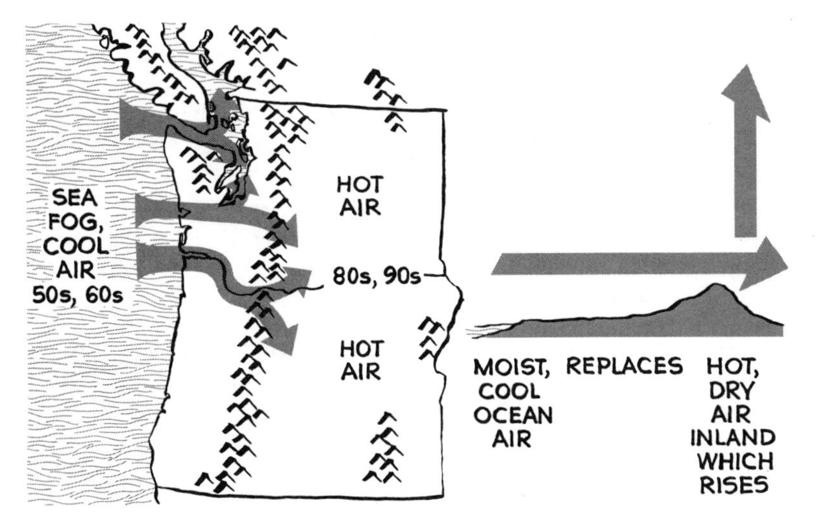


**Fig. 40.** *Formation of advection fog*

*Advection Fog Clues*

- Hot weather east of the Cascades in Washington, Oregon, and British Columbia (especially if that hot weather extends west of the Cascades).
- Cooling temperatures and fog moving northward along the coast.
- Air pressures at least 2 mb higher at Astoria or North Bend on the Oregon Coast than at Seattle.
- Strong westerly winds in the Strait of Juan de Fuca, the Columbia River Gorge, or the Fraser River Canyon, or northwesterly winds through Queen Charlotte Strait.
- Wind shift in Puget Sound/Willamette Valley or Strait of Georgia/Queen Charlottes from north or northeast to south or southeast.

Marine advisories and warnings can be valuable aids in gauging how thick and long-lived the fog or low-stratus layer is likely to be, since such alerts reflect the strength of the onshore push into the interior. In particular, look at the strength of winds through the Strait of Juan de Fuca or the Columbia Gorge. Remember, the wind direction must be generally from the west to drive the moist marine air inland.

*Marine Advisories/Advection Fog Guidelines*

| *Advisory* | *Precipitation* | *Burn-off Time* |
|---|---|---|
| None (10–20 kt) | Unlikely | Late morning/early afternoon |
| Small-craft (21–33 kt) | Drizzle possible | Late afternoon |
| Gale warning (34–47 kt) | Drizzle/rain | Next day |

The thickness of the advection fog or stratus layer offers clues as to how soon it may burn off. If the layer is more than 2,000 feet ( 600 m) thick, it may not clear at all on the first day. It may not clear until perhaps mid-afternoon on the second day, and then progressively earlier after that, assuming of course that the overall weather pattern doesn't change. Incidentally, southwesterly winds along the coast tend to thicken fog/stratus layers; northwesterly winds tend to thin them.

## Warm Frontal Fog

Our two previous types of fog offer the sailor or cruiser some glimmer of hope that visibility may improve later in the day or the option of traveling east to get out of the fog. The third type of fog offers no such hope or option beyond boarding a jet and getting far away, perhaps for a bareboat charter in the Caribbean. Warm frontal fog (fig. 41) covers a large area and is just as common east of the Cascades and the Coast Mountains as to the west.

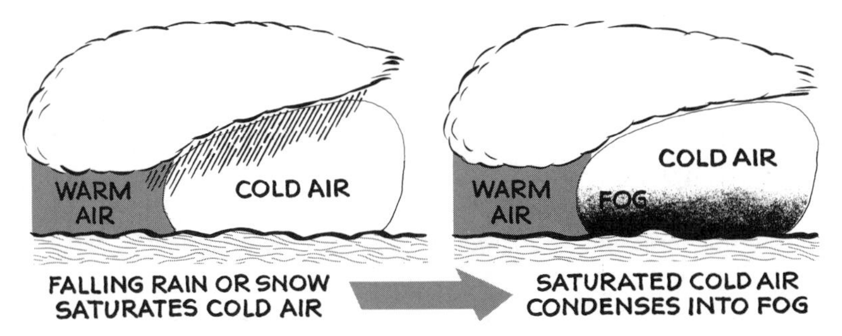
WARM AIR
COLD AIR
FALLING RAIN OR SNOW SATURATES COLD AIR
COLD AIR
WARM AIR
FOG
SATURATED COLD AIR CONDENSES INTO FOG

**Fig. 41.** *Formation of warm frontal fog*

Warm frontal fog is produced by precipitation falling from warm air aloft into colder air near the surface. The precipitation saturates the cold air, which then condenses into a thick layer of fog.

Unlike other varieties, warm frontal fog doesn't burn off. It disappears only after the surface warm front has arrived, bringing an end to the contrast between the cooler air near the ground and the warmer air running over it and an end to the precipitation formed by this process.

In the interiors of the United States and Canada, such fogs can persist for days during the winter. In the coastal regions west of the Cascades and the Coast Mountains, the duration is much shorter, simply because warm fronts tend to move through more quickly and because the contrast between the air masses is not as great.

### *Warm Frontal Fog Clues*

- Lowering, thickening stratus clouds.
- Light east to southeasterly winds.
- Steady precipitation.
- Small clouds with a shredded or torn appearance forming close to the ground.

## Sea Fog

From time to time, viewers ask, "What is that big gray blotch on the weather satellite pictures we see on television?" That blotch is most likely an extensive area of sea fog, frequently found in the North Pacific Ocean and the Bering Sea. In fact, sea fogs in these regions may thicken dramatically, reaching heights of 5,000 feet (about 1,500 meters).

Such fogs really are a variety of advection fog, because they're formed when relatively warm, moist air moves over cooler water. If the sea surface temperature is cooler than the dew point of the air, the moisture in the air condenses into fog, just as your breath does on a cold day. What may catch some skippers by surprise is that this fog may form even when winds are moderately strong, in contrast to radiation fog, which will form only when winds are light. Sea fog may persist in coastal areas and into the Strait of Juan de Fuca even with winds up to 30 knots, especially when winds are out of the west. This type of fog is most common from late spring to mid-autumn, when north to northwesterly winds have persisted for at least two to three days. Sea fog tends to reach a peak in late August.

*Sea Fog Clues*

- North to northwesterly winds along the coast.
- A minimum of two to three days of such winds.
- Initial wind velocities less than 25 knots.

When sea fog enters the Strait of Juan de Fuca during the prime months of July through October, it occasionally looks like a gray wall. Such sea fog is more likely to extend farther east along the Washington shoreline than along Vancouver Island. Even when most of the strait is a soupy mess, skies can often be clear north of Race Rocks.

### Steam Fog

Steam fog looks exactly like what its name implies: long tendrils of steam or smoke rising from the surface of the water. It's most common during the autumn and early winter months, when water temperatures are still relatively warm but air temperatures become quite cold. The cold air drains downslope toward and over the surface of a river, stream, lake, or bay. The water evaporating from the surface may saturate this much colder air almost on contact, rising with the air heated from the warmer water below. I enjoy sea kayaking or canoeing in such conditions, as the water is very still, the steam fog quite shallow, and the atmosphere both beautiful and eerie.

*Steam Fog Clues*

- Clearing overnight.
- Light winds.
- Air temperature cools below surface-water temperature.

## Localized Precipitation

### The Puget Sound Convergence Zone

While thunderstorms in the Pacific Northwest mountains are frequently the result of cool, unstable air moving over the warmer water of the Pacific Ocean, or from rapid lifting against mountains, there are other causes. One is the Puget Sound Convergence Zone.

This convergence zone (fig. 42) can produce locally heavy precipi-

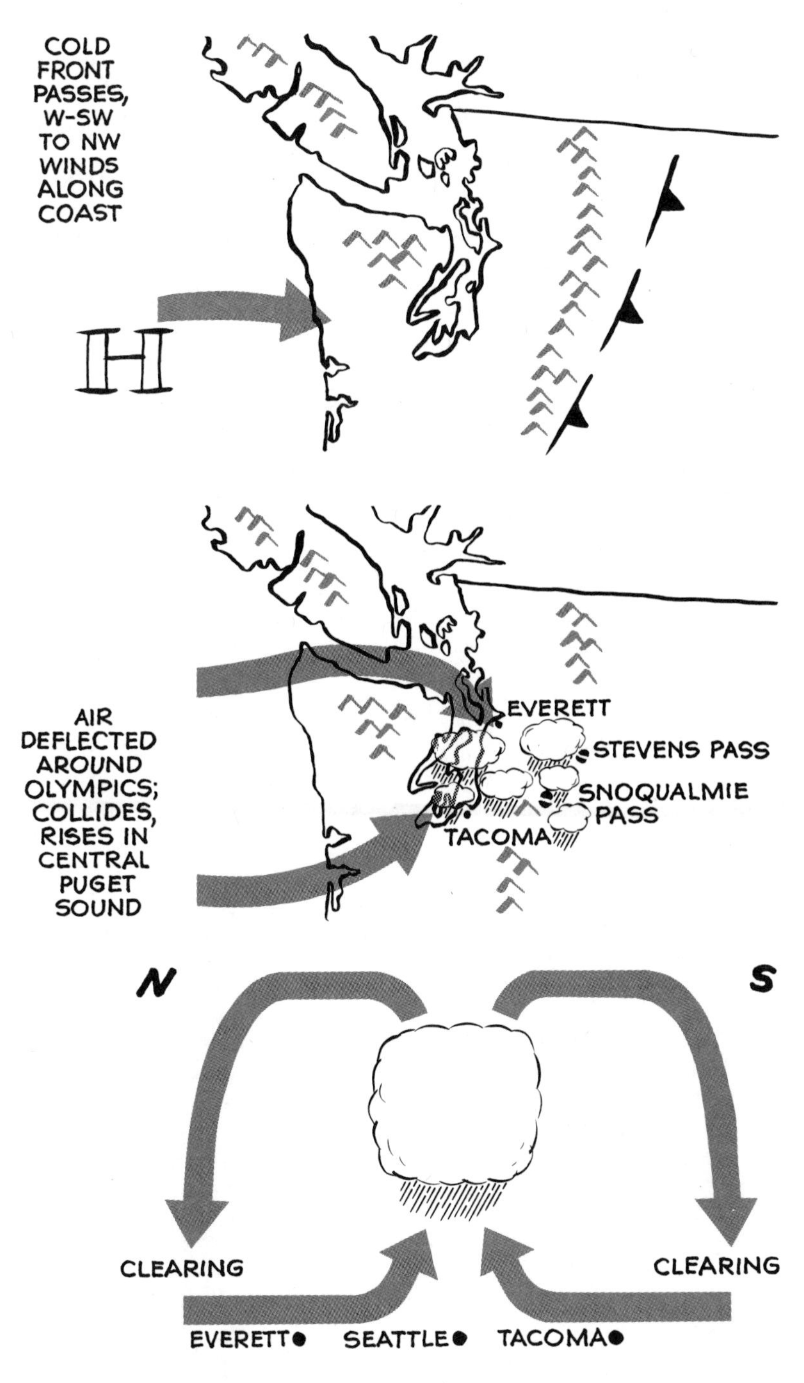


**Fig. 42.** *Puget Sound convergence zone*

tation while other areas just a few miles to the north or south may be enjoying sunshine. The zone develops after the passage of a cold front as high pressure builds into the coast, producing west–southwest to northwesterly winds.

When the onshore flow of air runs into the Olympics, it splits, some flowing through the Strait of Juan de Fuca to the north, some flowing through the Chehalis Gap to the south. The Cascades present an almost insurmountable barrier to the east, so some of the air moving through the strait is forced south into Puget Sound while some of the air moving through the Chehalis Gap is forced north. The two opposing currents collide, forcing some of that air to rise.

This convergence zone ranges from Everett to Tacoma, tending to move south during the afternoon and the early-evening hours. A weak convergence zone may produce only thicker clouds within central Puget Sound. A stronger zone may produce locally heavy rain or snow and possibly even thundershowers.

*Convergence Zone Clues*

- Cold-front passage.
- West–southwest to northwesterly winds along the coast.
- North to northwesterly winds near Everett.
- Southerly winds to the south of Seattle.

Because the rising air within the convergence zone sinks to the north and to the south, there's occasionally clearing in those areas and certainly reduced precipitation. The best plan if a convergence zone is active or likely to be active is to head south of Tacoma or north of Everett.

## Icing

Icing may seem an unlikely problem in the Pacific Northwest, a hazard to be expected farther north, in Alaskan waters. While it's true that subfreezing temperatures are relatively rare along the coast, arctic fronts do push through. When they do, the potential for dangerous icing can be high. No vessel, however large or small, is exempt from this risk. Small craft coated with ice become top heavy and, if winds or waves are sufficiently strong, can abruptly capsize. Even a light coating of ice can make decks slippery.

Arctic outbreaks are the usual source of icing conditions, a southward push of frigid air from the interior of British Columbia, the Yukon, and Alaska. The air jets out from river valleys and passes, blasting coastal and interior waters with subfreezing air carried by strong, gusty winds that in turn create rough seas. The combination of strong winds and cold air usually begins to produce freezing sea spray once temperatures drop below 28 degrees Fahrenheit (–2° C), with the heaviest icing found once temperatures drop into the teens, as shown in the table that follows. The ranges shown are approximate, and icing severity increases with higher wind speeds. The definitions of icing types are given in inches of accumulation per hour.

| *Icing Type* | *Wind Speed (knots)* | *Air Temperature (°F/°C)* |
|---|---|---|
| Light (less than .25" per hour) | 12–20 | 20 to 28/–6 to –2 |
| Moderate (.25–.75" per hour) | 20–30 | 15 to 20/–10 to –6 |
| Heavy (more than .75" per hour) | Above 30 | Colder than 15/ colder than –10 |

### The Silver Thaw

When a dome of cold air flows from the east through the mountain passes, it thrusts under the warm, moist air engulfing the region to the west of the Cascades. Occasionally this warm air is producing rain, and as it falls into the shallow layer of cold air below, that rain freezes, glazing over everything.

Such a pattern is often called a silver thaw, because the coating of ice gives trees, rocks, roads, and boat decks a silvery-white appearance. Such patterns are most common along the Fraser and Columbia rivers during the months of December, January, and February. The worst cases occur when the Pacific Northwest has been locked in subfreezing weather and a low-pressure system is approaching from the west, bringing warmer air and the promise of rain. Eventually the warm air replaces the cold air flowing from the east, and the weather really does begin to thaw.

## Thunderstorms

Cold fronts in much of the United States and Canada are generally followed by clearing and an end to precipitation. But western Oregon, Washington, and British Columbia operate, as usual, by a different set of rules. Here, the heaviest precipitation may actually follow cold fronts.

The air that moves into coastal areas along a cold front has been warmed by the relatively mild ocean temperatures. By the time it arrives, there's often little difference in temperature between the once-cold air mass that moved over the ocean and the warmer air mass over land. Without a strong temperature contrast, the collision between the two air masses is more of a gentle nudge, and nudges don't build up towering cumulus clouds. However, once the cold front moves through, if the winds aloft shift to a more northwesterly direction, much colder air originating over the interior of Alaska and the Yukon streams over the warmer Pacific Ocean. This time, the distance the cold air moves over the ocean is much shorter, and two major steps occur.

First, the cold air is heated by the ocean, which may be anywhere from 10 to 80 degrees Fahrenheit warmer than the air above. This is a prime example of unstable air. Think of a pot of water that's being heated on the stove. The water on the bottom is expanding through heating. Small bubbles blossom into bigger bubbles, which need more room, which continue to expand by rising. Unstable air behaves the same way.

Second, the Pacific Ocean supplies something else the cold air from the north doesn't have: moisture. The water vapor that rises into the colder air aloft cools and condenses into liquid water droplets. These droplets may eventually develop into towering cumulus or cumulonimbus clouds.

The upward nudge that starts the whole process may simply come from the air moving over the warmer ocean. It may also come from the air moving over the sun-warmed land of coastal Washington, Oregon, or British Columbia; even partial breaks in the cloud cover sufficiently warm the land to provide that boost.

And there's yet another agent that can help provide this upward shove: mountains. When the unstable air blows up against the mountains, they act as barriers, and the air has but one direction to go: up. As the air rises, there is often sufficient cooling to jump-start the condensation that produces clouds and precipitation. The name for precipitation generated in this manner is orographic precipitation.

The lifting of air over mountains with even an average wind is often 50 to 100 times as great as the lifting along a cold front. This creates tremendous amounts of precipitation on the windward side of coastal mountains. In fact, roughly 60 percent of the precipitation falling on them is created after, not during, frontal passage. That's why most thunderstorms along the Pacific Northwest coast occur well after cold-front passage. Such thunderstorms are nothing to trifle with (fig. 43), even if relatively puny when compared to the monsters east of the Rockies. An early spring thunderstorm can release up to 125 million gallons of water and produce winds gusting 40 to 80 miles per hour.

The greatest number of thunderstorm fatalities on the water come from lightning. The exact cause of lightning still eludes scientists. However, researchers know the initial discharge stroke, or "leader," moves almost invisibly from the cloud toward the earth. No thicker than a pencil, this discharge stroke attracts electrical charges on the ground, which connect with the leader and race upward toward the cloud in the much larger return stroke, which is what we see. A single lightning bolt may have an electrical potential of 125 million volts, heating up to 50,000 degrees Fahrenheit—roughly five times the temperature of the sun's surface! Two hundred people die from lightning strikes in the United States each year, many out on the water and occasionally in the Pacific Northwest. With a few precautions, most such accidents could be avoided.

*If Thunderstorms Are Forecast*

- First watch small cumulus for strong upward growth.
- Keep track of weather reports hourly.
- Listen to AM radio broadcasts for strong static interference, which may come from lightning.
- Consider altering course to keep you near safe anchorages.

*If You Spot Thunderstorms or Hear Thunder*

- Seek safe anchorage immediately if possible.
- If safe anchorage is not possible, seek safe shelter inside your craft.
- Prepare for gusty winds in rough water.
- Gauge the movement of the thunderstorms.

As for that last point, how do you gauge the movement of thunderstorms? It's easy if you have a watch. The moment you see lightning, start timing. Stop timing once you hear thunder. Divide the number of seconds by 5. The result is the distance of the thunderstorm from you in miles. Continue to time for lightning and thunder discharges to judge whether the thunderstorm is approaching, remaining stationary, or receding. If the time interval between the lightning and thunder is decreasing, the thunderstorm is approaching. If the interval is increasing, the thunderstorm is moving away.

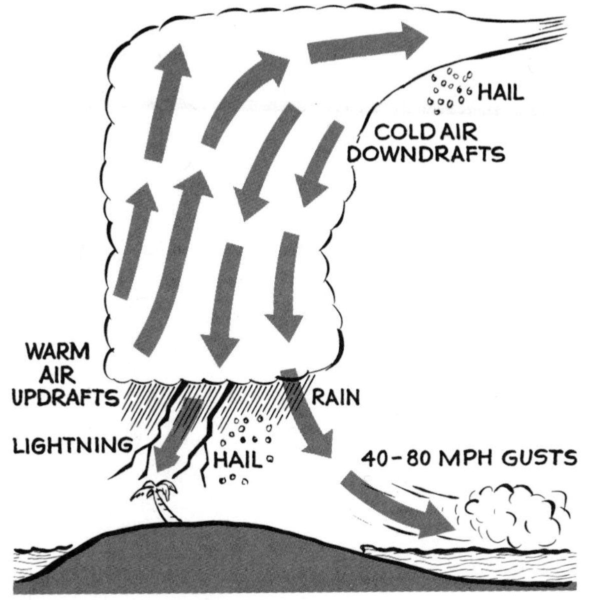


**Fig. 43.** *Thunderstorm hazards*

The above technique works because the speed of light is very fast, 186,000 miles per second (essentially instantaneous). The sound of thunder is much slower, moving about 1 mile every 5 seconds. If you don't hear thunder, the storm is probably at least 15 miles away, as thunder can be difficult to hear at that distance. I've used this technique successfully both sailing and kayaking. During a kayak trip from Seattle to Port Townsend, thunderstorms started booming away to the south of our position near Point No Point. Using this technique repeatedly, I determined the action was remaining stationary to the south. We continued paddling to the north, without a hitch.

Thunderstorms are most common along the Pacific Northwest coast after cold-front passage during the mid-autumn, winter, and spring months. However, thunderstorms occasionally move north in the summer from California, first into Oregon's Willamette Valley and then into Puget Sound, the San Juans, and the Gulf Islands. Such thunderstorms are associated with the end stage of a hot spell just before the onshore push we discussed earlier. The onshore push of marine air following such thunderstorms is just about always guaranteed to be strong.

Thunderstorms are also associated with occluded fronts (see the discussion of occluded fronts near the end of chapter 2). Such thunderstorms are most likely when the low-pressure system generating the occluded front has a central pressure of about 28.65 inches of mercury (about 975 mb) or less. Expect strong gusty winds of gale force or greater. Small craft have no business out in such conditions. Follow the guidelines listed earlier for watching for thunderstorms and taking action if you spot them. Keep your eyes on the sky and your ears on NOAA or Canadian weather radio to reduce the odds of such exposure.

CHAPTER 6

# A Forecaster Afloat: Using the Tools of the Trade

*Red Sky at Night, Sailors Delight;*
*Red Sky in Morning, Sailor take Warning.*

Unknown

Since the first voyagers left the shore behind in search of food, treasure, or adventure, there's been an ever-increasing attempt to understand and anticipate changing weather. This age of computers, conventional and satellite-directed broadcasting, and cellular communication has opened new weather tools to the mariner. Over the course of the past five chapters, we've discussed guidelines that can assist sailors, powerboat skippers, kayakers, and windsurfers in enjoying their time on the water safely. These guidelines are worthless, however, without a foundation of good, current weather information.

As a flight instructor, I teach my students to obtain thorough weather briefings and to analyze the effects of the forecasts and current weather information on their planned flights prior to takeoff and while en route. This is no less important for someone heading out onto the water. If the weather deteriorates, a pilot can move away at speeds of 100 miles per hour or more. The speed of a mariner is considerably slower, of course, increasing the hazard of exposure.

There are many good sources of weather information, both at dockside and while under way. This chapter reviews those sources. The next chapter suggests ways to analyze the information for safe and effective decision-making.

## Weather Information Sources for Use at Home

### Print/Broadcast

Whether you're planning a day sail or a circumnavigation of Vancouver Island, the sources used for gathering weather information are often the same; only the depth of research changes. Television, radio, and newspapers are the most widely used sources of weather information by the public. Unfortunately, the most widely used sources aren't necessarily the best.

Newspaper weather information, although offering you the option of rereading and digesting the data at your leisure, is dated and occasionally obsolete by the time it reaches your doorstep or the newsstand. An increasing number of dailies farm out their weather pages to private forecasting firms (located well outside the Northwest) whose forecasters have little understanding of this region's weather patterns. Check to make certain your paper uses the National Weather Service or Environment Canada forecast, or a forecast made by a local consulting meteorologist.

Radio and television weather information is usually current, but often aggravatingly brief—and its reliability and timeliness depend heavily upon the expertise of the person delivering the forecast. Before relying on a broadcast weather presentation, be certain the weathercaster is a meteorologist with a degree in the field. That's your best guarantee that the information will have been updated to the moment, accounting for local variations in the forecast area. Too many stations foist off an announcer on the public with at best scant training in the field, often with the excuse, "Aw, you don't need all that theoretical stuff to forecast!" Just as it's helpful for a physician to have completed professional training before going to work on you, a professional degree is important for a weathercaster. The weather services of both the United States and Canada require it and so should you as a consumer of weather information.

## Telephone Weather Sources

Several good sources of weather information are available over the telephone. The U.S. Coast Guard, the National Weather Service, and Environment Canada's Pacific Weather Centre offer recorded summaries of marine weather conditions and forecasts on local numbers for the following areas:

Astoria: (503) 861-2722
Cape Disappointment: (206) 642-3565
Neah Bay: (206) 645-2301
Port Angeles: (206) 457-6533
Seattle: (206) 526-6087
Vancouver, B.C.: (604) 270-7411

Both the Federal Aviation Administration and Transport Canada offer recorded telephone weather information. Although these recordings are intended for pilots and obviously won't include information such as sea state, the summaries of current weather observations and forecasts are helpful. Callers with a touch-tone telephone can access a menu of different observations and forecasts.

FAA: (800) WXBRIEF
Transport Canada/Vancouver: (604) 273-1151

Windsurfers may be particularly interested in a service called Windsight, based in the Portland area, which specializes in wind forecasts, particularly in and around the Columbia River Gorge. The company offers a menu of forecasts and observations accessible by touch-tone telephone for a fee based on the time you're connected. Call (800) 934-2278.

Some broadcasting stations also offer recorded weather information. My station, KING -TV in Seattle, offers such a service at (206) 448-3649.

## Computer and Facsimile Data Sources

The computer has moved from the workplace into the home as an almost indispensable tool. Just as it can crunch through large volumes of data, quickly produce eye-catching graphics outlining trends and budgets, act as a tutor, and connect users to vast information sources, it now offers skippers the opportunity to tap into weather data on demand.

Both the CompuServe and Prodigy services offer general weather information. Other computer services offer highly specific weather data and forecasts that can be tailored, for a fee, to your planned voyage. Some, like WSI of New Bedford, Massachusetts, have even developed a Windows-like menu system that is a breeze to use. The beauty of such sources is that the information is always there; users can get it when they need it or want it, not just when someone else decides to make it available. It's an excellent way to get a thorough picture of present and future weather before a trip.

The fax machine that allows you to bypass mail service also can serve as a source for current weather data, maps, and forecasts. Several services allow users to order a series of weather maps, observations, and even satellite pictures simply by calling a fax number and entering a password and a series of request codes. These fax services are easy to use and relatively inexpensive. Ordering a satellite image, surface and upper air maps, forecast map, and current weather observations may cost as little as $10—and it can be done from your office fax machine before you leave for the marina!

Expect the area of computer and facsimile weather data to change rapidly with expanding services. Such rapid change makes it impractical to attempt to offer a definitive listing of such services. See Appendix 4, which lists a few such services that I've investigated and found worth trying.

# Weather Data Sources for Use Aboard Your Boat

## Barometer

A barometer is one tool you shouldn't be without in your home or on your boat. It offers valuable information and guidance on the speed at which a weather disturbance is approaching or at which a high is building in. If you purchase one of the fancy brass versions, choose one with a resettable guide, which allows you to quickly see how much the pressure has risen or fallen since you last checked. If you don't own a boat, but charter or crew on other vessels, consider purchasing a pocket-size digital barometer. It will give you not only the current pressure reading but also the pressure trends over various time intervals. Such digital barometers also make a better choice than the fancy brass versions for a small vessel. (The digital pocket barometers can be found at most mountain outfitters. Ask a climbing friend for recommendations.)

Don't be lured into spending a bundle on a barometer, believing this will buy you the best instrument. Captain's Nautical Supplies in Seattle, which not only sells but also repairs barometers, tested about thirty different barometers for accuracy, sensitivity, and temperature compensation. Some expensive models performed very well, but many inexpensive models also performed very well—another example of not judging the character of an instrument by its casing. The test results are available from Captain's.

Once you purchase a barometer, a few corrections may be necessary to make your instrument accurate. Errors can result from the variations in gravity found at different latitudes, from differing heights above sea level, nonstandard temperatures, and index errors related to the instrument's design and manufacture. Most skippers purchase aneroid barometers instead of the older mercury version. Aneroid barometers usually don't need to be corrected for gravity or temperature variations and, unless they are used on inland lakes or rivers, the correction for altitude will usually be negligible. The necessary correction for altitude can be made by finding the elevation of the lake or river, then referring to the *Smithsonian Meteorological Tables,* found in most libraries. The correction for index errors is necessary because of changes in the elasticity (think of it as stretchability) of the metal surrounding the vacuum chamber. Just call the local National Weather Service office for a correct reading. It's wise to do this once each year.

Incidentally, a pressure rise is no guarantee of wonderful weather here in the Pacific Northwest. If you get reports of a high within 600 to 700 miles (960–1,120 km) of the coast with a low in eastern Washington, eastern

Oregon, or southeastern British Columbia, and you note a 3-hour pressure rise of 2 to 4 millibars, then gale-force winds are possible; that is, wind speeds exceeding 33 knots. Given the channeling effect of passes and gaps in the terrain, speeds of 60 knots in the mountains can develop. Consider waiting until the high has had an opportunity to build into the Northwest; that usually happens within 12 to 24 hours.

## Anemometer and Wind Vane

An anemometer and wind vane are also good investments for your boat. Wind vanes offer the obvious benefit of a direct readout of wind direction and permit you to monitor changes in wind direction. Anemometers allow you to keep track of wind speed and, of particular importance, changes in wind speed. Remember that when both instruments are read aboard a moving boat, there will be some errors in the wind direction and speed depending upon the direction and speed of your vessel. Just as a barometer requires occasional calibration, wind vanes and anemometers require occasional maintenance. The manuals provided with the instruments explain the care needed. Write down those items on your normal maintenance schedule. It goes without saying that if such instruments are worth purchasing, it's worth spending enough to buy reliable instruments. Purchase them from a ship chandlery you trust.

Remember the following points when monitoring your weather vane:

- A shift from north or northeasterly winds to southeasterlies generally signals the approach of a low.
- A shift from southeast to southwest usually signals the passage of a cold front.
- A shift from southwest to north usually indicates the building of high pressure.

## Shipboard Weather Radio

A radio capable of receiving weather broadcasts from the National Weather Service (often called NOAA weather radio) or Environment Canada should be considered mandatory safety gear aboard every boat. These broadcasts offer regular updates on weather throughout the region as well as forecasts and warnings of impending severe conditions. You can purchase either a dedicated unit that receives only weather broadcasts or a standard marine VHF radio that receives the frequencies used by the weather broadcasts (typically channels one and two).

The dedicated weather radio-frequency units can be purchased for less than $100 (do avoid the bargain-basement versions; their longevity only slightly exceeds that of the trinkets offered in fast-food meals for children). I prefer a standard marine VHF radio because of its flexibility;

it can receive other frequencies and can also be used as a radiotelephone. This flexibility gives you the option of radioing for assistance if you get in trouble.

Portable VHF radios are available that are no larger than a cellular phone, and they are a wise choice for either small-boat owners or skippers who charter vessels for their trips. If you purchase a portable VHF radio, do buy an extra battery pack. Listening usually drains the battery less than transmitting does, so checking the National Weather Service (NOAA weather radio) or Environment Canada weather broadcasts shouldn't require a battery change over the course of a two- or three-day trip. The battery may even last as long as a week.

## Shipboard NAVTEX

Distance and terrain can block or degrade radio transmissions. Both are common elements that can reduce the utility of weather radio broadcasts for skippers cruising or sailing in the Pacific Northwest. The skipper on an extended trip may also wish to obtain weather data beyond the immediate area, particularly if plans call for cruising in exposed waters, such as along the coast. For this purpose, two pieces of equipment are worth considering: a NAVTEX receiver and printer, and a weather facsimile receiver and printer (described in the next section).

NAVTEX is a marine radio warning system utilizing a single side-band radio and a small printer. It transmits on the international standard medium frequency of 518 kHz. Urgent messages are broadcast upon receipt, while routine messages and transmissions are broadcast four times a day. Information available includes notices to mariners, gale and storm warnings, and offshore forecasts. Ship chandleries sell either dedicated NAVTEX receivers with printers (at the time this book was prepared, prices averaged about $1,200) or printers that can be attached to an existing single side-band radio. The beauty of this system is that you don't have to be listening to the radio to receive messages and warnings, because they're printed out for later reference. The system is undergoing revision, but transmitting stations are either in operation or planned for Astoria, Oregon, and for Vancouver, British Columbia. Contact the U.S. Coast Guard for an update.

## Shipboard Weather Facsimile

Weather facsimile receiver/printers turn your vessel into a floating forecast center in which you can receive satellite images along with weather analysis and forecast maps. The maps give an excellent picture of current weather in your area as well as a better appreciation for the overall pattern that could impact your plans a few days in the future. Probably the best and most reliable manufacturer of such units is Alden

Electronics. The units are available at larger marine electronics outlets, with prices ranging between $3,000 and $3,500.

If plans call for some open-water cruising or racing, or even extended cruising where you need a complete picture of the weather, weather facsimile should be considered an important piece of safety equipment. The products available include satellite images; surface analysis maps; maps showing sea state, wave height, and the upper air; and maps projecting future positions of highs, lows, fronts, and precipitation.

For cruising in the Northwest, skippers will be tuning into stations transmitting from one of two locations. All stations transmit at a power of 10 kilowatts.

| Stations | Frequency | Operation | Power |
|---|---|---|---|
| Esquimalt, B.C. | 4266.1 kHz | Continuous | 10kw |
| | 6454.1 kHz | Continuous | 10kw |
| | 12751.1 kHz | Continuous | 10kw |
| San Francisco | 4346 kHz | Night | 10kw |
| | 8682 kHz | Continuous | 10kw |
| | 12730 kHz | Continuous | 10kw |
| | 17151.2 kHz | Continuous | 10kw |
| | 22528.9 kHz | Day | 10kw |

## Interpreting Shipboard Facsimile Charts

A common question from sailors or powerboat skippers attending my weather seminars is, "I know how to tune in and receive the various weather maps, but how do I interpret them once I have them?" That's the purpose of the following section: to show some of the key maps available through your on-board facsimile receiver/printer and to discuss how to use them.

First, a note about time. Although the date marked on the maps and charts is straightforward to read, the time given is Universal Time (formerly called Greenwich Mean Time). During those months when the Pacific Northwest is using daylight savings time, the Northwest is seven hours behind the time indicated; during standard time, the Northwest is eight hours behind. A chart labeled 1200 during daylight savings time would therefore be showing conditions at 5:00 A.M. local time (seven hours earlier). A chart labeled 0000 would be showing conditions at 5:00 P.M. (seven hours earlier). Remember, 0000 is equal to 2400 hours.

### Surface Analysis Map

The surface analysis map is the one most people visualize when they think of weather maps. It gives the locations of surface high- and low-pressure centers, troughs, fronts, and weather observations. As shown in the key for figure 44, the location of surface highs and lows is self-explanatory; the center of a high is marked with an H; the center of a low is marked with an L. The number next to the letter is the central pressure in millibars.

The lines snaking around the highs and lows are called isobars; they connect areas of equal pressure. Those pressures are marked along each line in millibars, though only two digits are generally given. If the number is fifty or greater, the missing digit will be a nine. (For example, add a 9 in front of the given number of 70, for a pressure of 970 millibars.) If the number is less than fifty, the missing number will usually be a ten. (For example, add a 10 in front of the given number of 30, for a pressure of 1030 millibars.) The closer the isobars are to each other, the faster the winds will blow; conversely, the farther they are spaced apart, the lighter the winds will be. The winds typically blow across the isobars, toward lower pressure, at about a 10- to 15-degree angle over open ocean, at about a 20- to 30-degree angle over water surrounded by fairly flat land (such as Puget Sound), and at up to a 90-degree angle on waters surrounded by steep terrain (such as the Columbia River Gorge, Desolation Sound, and Princess Louisa Inlet).

Cold fronts are marked by lines with a series of triangles, warm fronts by lines with a series of half circles, and occluded fronts by lines with alternating half circles and triangles. The marks on the line point in the direction the front is moving. From time to time, there will be an arrow near the front with a number, showing the direction the front is moving and its approximate speed in knots. There won't be such direction/speed indicators near a line with triangles pointing one direction and half circles the other, because this is a stationary front.

Finally, the surface analysis map has a series of circles representing observing stations. As the key in figure 44 indicates, the circle represents cloud cover. An open circle represents clear sky, a partially filled circle is partly cloudy skies, and a filled circle is cloudy skies. The line attached to the circle points to the direction the wind is coming from, and the "barbs" attached to the line indicate wind speed. Each full barb represents 10 knots, a half barb 5 knots, and a pennant 50 knots. The other numbers are easily understood by simply examining the key below the sample map.

To summarize, key elements to examine in a surface analysis map include:

- Positions/pressures of highs and lows.
- Closeness of isobars.
- Positions and types of fronts.
- Station observations.
- Changes since the previous surface analysis map.

I sometimes mark prior pressures and positions of lows, highs, and fronts on the most recent map, with a notation of the time of those prior

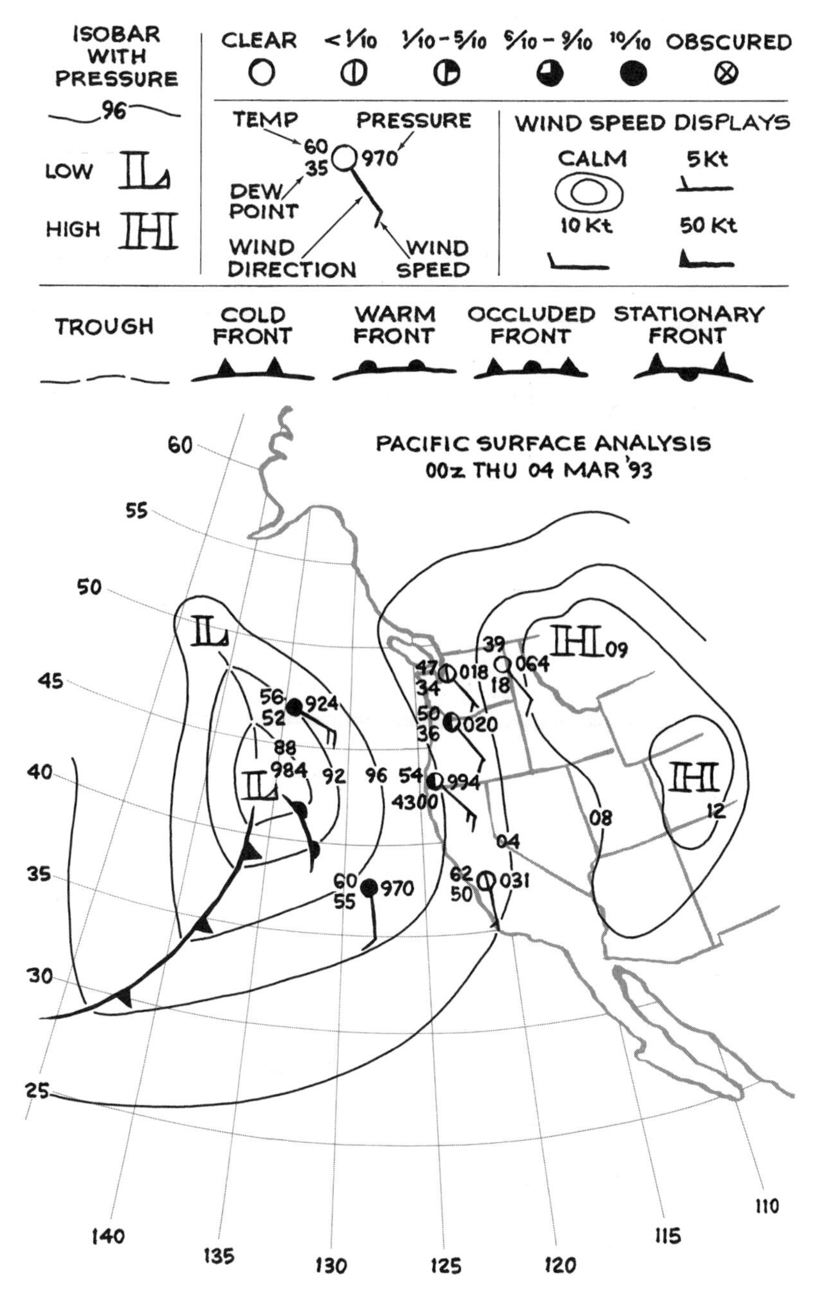

Fig. 44. *Surface analysis map*

positions. This helps in analyzing intensity, direction, and speed of movement, as well as changes in observed weather. Historical knowledge is helpful in weather forecasting, as well as in politics.

## Upper-Air Analysis Map

Surface weather analysis maps show where weather systems are and where they have been going. Upper-air maps help indicate where such systems are likely to go in the future. They indicate wind direction and velocity at various altitudes, and occasionally temperatures and moisture content.

The upper-air analysis map shows the location of high- and low-pressure centers. But they don't tend to be closed off. Those areas where the contour lines are rising toward the north over a high are called ridges; those dipping toward the equator under a low are called troughs. There are several different upper-air analysis maps. The ones most commonly transmitted over marine radiofacsimile units are called 850-millibar and 500-millibar maps.

On surface analysis maps, the contour lines, called isobars, connect places of equal air pressure. On upper-air maps, the contour lines connect places where the reference pressure is found at the same altitude. For example, on the sample 500-millibar map shown in figure 45, the contour line marked 570 indicates that the pressure of 500 millibars is found at 5,700 meters above sea level (roughly 18,000 feet). The 500-millibar upper-air analysis map is roughly centered at 18,000 feet, while the 850-millibar upper air analysis map is roughly centered at 5,000 feet. This information, by the way, is generally obtained through soundings of the upper atmosphere using weather balloons, although satellites can now obtain some of the same data.

The handy thing about these upper-air analysis maps, especially the 500-millibar map, is that the contours tend to parallel the wind at those altitudes. The force of friction, which causes air to move from high to low pressure at and near the surface, is for all practical purposes absent in the upper atmosphere (at least as far as meteorologists are concerned). So by examining a 500-millibar map, you can see where the wind will steer surface weather disturbances, and even how fast they'll tend to move. Surface disturbances tend to move at half the speed of the wind at 500 millibars.

The other useful aspect of 500-millibar maps is that they tend to show where surface lows are most likely to form, as well as surface highs. As shown in the key for figure 45, surface lows are usually found, or are most likely to develop, from the bottom of the trough downwind to the top of the ridge. Surface highs are usually found, or are most likely to develop, from the top of the ridge downwind to the base of the trough. If a split exists in the 500-millibar air flow, surface disturbances usually will weaken or entering the split. Such knowledge is handy for skippers in here trouble may develop.

arize, the following are key questions to answer in analyzing nap, particularly a 500-millibar map.

re the ridges and troughs?
re the surface lows on the map?
the upper-air winds show concerning where the surface be steered?
e splits that will tear apart surface lows?
the pattern changed since the previous upper-air map?
night a future surface low develop, and is it near you?

## nages

opment of weather satellites has resulted in huge advances understand and forecast the workings of our atmosphere. here are these benefits as great as along the West Coast, ation on weather systems developing over the ocean had en limited to a few ship reports. Satellite images are an uable tool in analyzing and anticipating weather patterns thin or approaching the Pacific Northwest, where observing ue to be sparse. A single satellite picture really can be worth ords.

ey patterns are helpful to recognize in satellite images (fig. f clouds, resembling a big white caterpillar, generally marks ont (A). A bend that develops (B) typically marks the birth w, with associated cold and warm fronts. The approach of aped cloud (C) often causes the infant low to intensify tern resembling that in (D) marks a mature low, which has nd occluded fronts. The spiral surrounds the surface low, ea of steep pressure gradients, which in turn often generate nd rough seas. A field of cumulus clouds (E) may remain r cumulus or may develop into cumulonimbus clouds. If a ed cloud (C) approaches, the cumulus clouds may organize howers. Finally, a blanketlike appearance of the clouds of marine stratus or fog. Satellite interpretation is a highly kill, built over years of training and practice, but devel- reciation of the basic elements will greatly improve your

## Prognostic) Charts

charts, commonly called prognostic charts (meteorologists her scientist—we like to use fancy words, too!), depict future



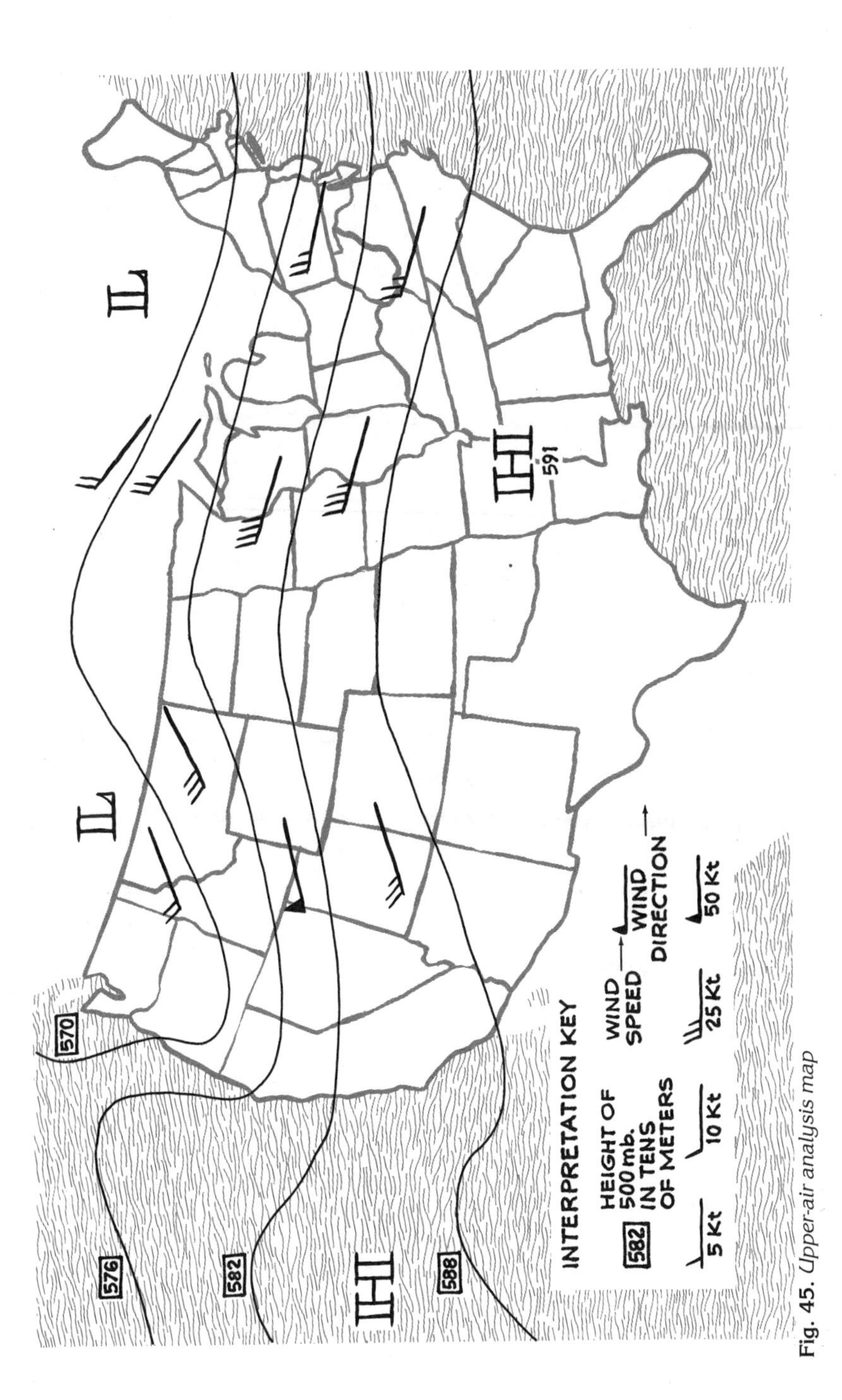


Fig. 45. *Upper-air analysis map*

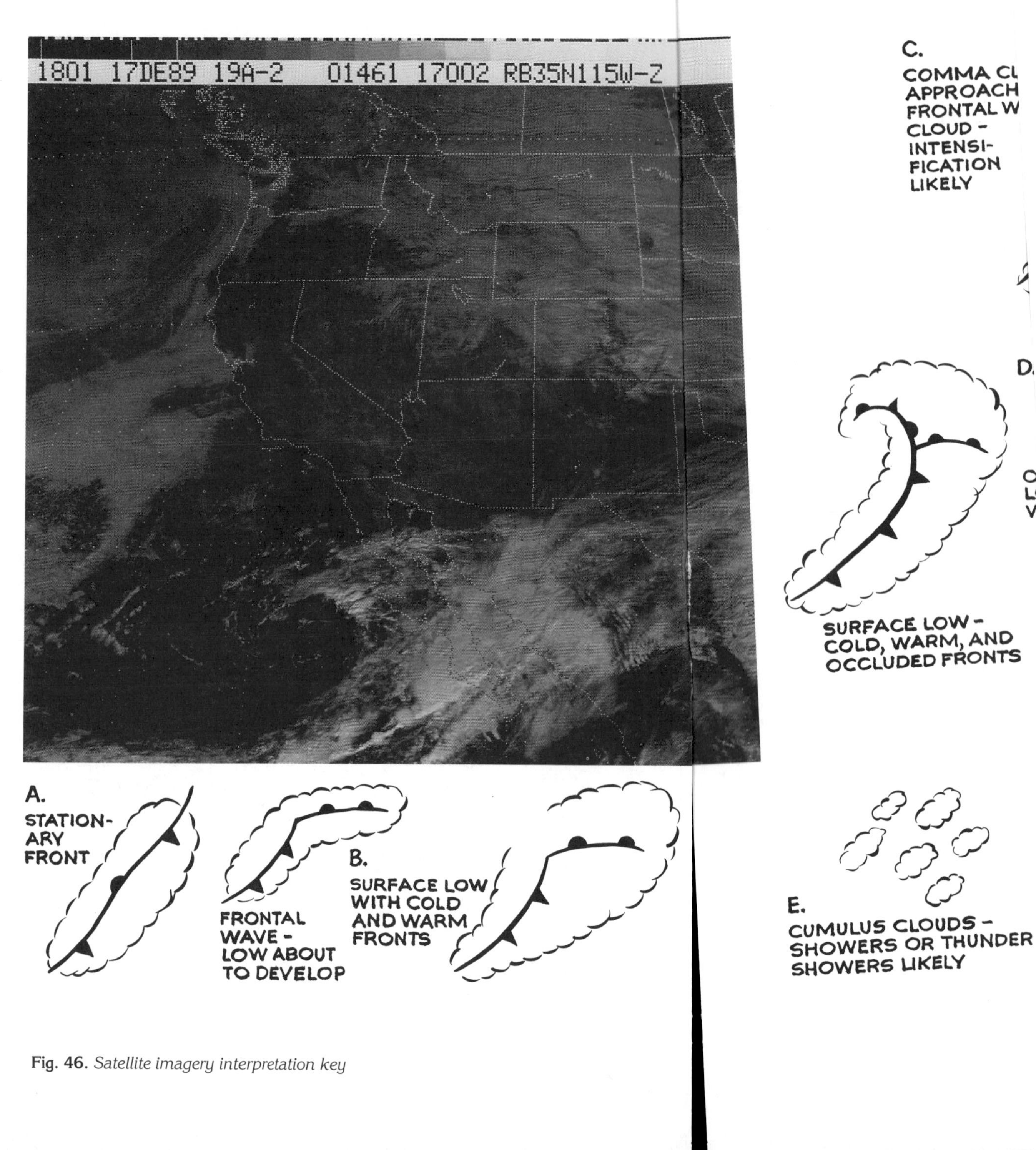


Fig. 46. *Satellite imagery interpretation key*

dissipate after
anticipating w
To summ
an upper-air r

- Where a
- Where a
- What do
  lows wil
- Are ther
- How has
- Where r

## Satellite Im

The devel
in our ability t
Perhaps nowh
where informa
previously be
especially val
developing wi
stations conti
a thousand w
Several k
46). A band o
a stationary fr
of a surface lo
a comma-sha
rapidly. A pat
cold, warm, a
marking an ar
strong winds
as fair-weathe
comma-shap
into a line of
(F) is typical
specialized s
oping an app
weather eye.

## Forecast (

Forecast
are like any ot

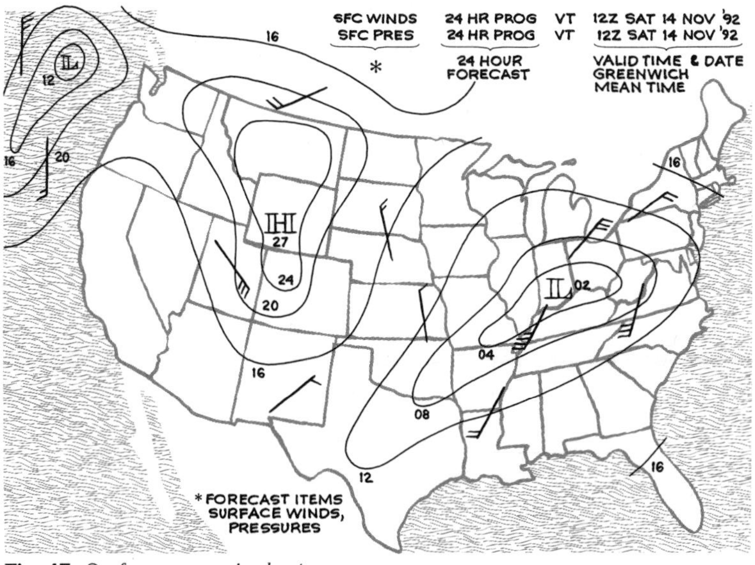


**Fig. 47.** *Surface prognosis chart*

expected conditions at the surface or in the upper atmosphere. They show the same types of information found on surface or upper-air maps that analyze current conditions—but the information is a projection of future conditions for some specified time. A 24-hour surface prognostic chart (fig. 47) illustrates expected placements and intensities of surface lows and highs 24 hours from the time it was issued. Simply compare such charts to present conditions to get a sense of how the conditions are expected to change. You can also keep such charts to check against future conditions to see how close the forecast comes to what actually transpires. If it is too far off, it's a tough forecast that has meteorologists guessing, and you should be extra cautious in your planning.

Other weather-information products are available, in addition to the ones we've discussed in this chapter, but few of these are in general use in the Pacific Northwest. You should also realize that the maps available over your marine weather radiofacsimile are only a sample of those used by professional meteorologists in developing a forecast. Use the maps as a supplement to such forecasts or when information is lacking. Continue to obtain and pay close attention to forecasts issued by the National Weather Service or Environment Canada.

CHAPTER 7

# Creating Your Own Solid, Confidence-Building Weather Briefings

*When the Wind is in the East,*
*'Tis neither good for man nor beast;*
*But When the Wind is in the West,*
*Then it is the very Best.*

Izaak Walton
*The Compleat Angler*

The mariner has a rich supply of sources for weather information. We covered many of them in the last chapter. But information by itself is of limited use; information gathered with a purpose is of great value.

Given all these information sources, the key is to use them well. To best assess the suitability of weather on the water, follow a plan. The following checklist is one such plan. Use it, modify it, or develop one better suited to your needs. The key is to organize your weather-information gathering.

*Marine Weather Checklist*

**Planned Trip:**______________________________________________

______________________________________________________________

| **Checklist** | **Data Sources** |
|---|---|
| ___Synopsis | Weather Radio Frequencies__________ |
| ___Warnings/Watches | Telephone Weather # __________ |
| ___Zone Forecasts | ____________________ |
| ___Marine Forecasts | Computer Data Source __________ |
| ___Offshore/Bar Conditions | Modem #__________________ |
| ___Extended Outlook | User ID __________________ |

___Local Observations Password ____________________

___Barometer

___Notebook/Weather Logbook

**Weather Synopsis:** ________________________________________

________________________________________________________

________________________________________________________

**Warnings/Watches:** ________________________________________

________________________________________________________

________________________________________________________

**Zone Forecast(s)**

| | Date:________ | Date:________ | Date:________ |
|---|---|---|---|
| Sky/Weather | ____________ | ____________ | ____________ |
| Temperatures | ____________ | ____________ | ____________ |
| Winds | ____________ | ____________ | ____________ |

**Extended Outlook** from ____________ to ______________:

________________________________________________________

**Marine Outlook**

Location:_________________ Wind/Waves/Swell:_________________

Location:_________________ Wind/Waves/Swell:_________________

Location:_________________ Wind/Waves/Swell:_________________

Location:_________________ Wind/Waves/Swell:_________________

**Weather Observations**

| Location | Sky/Weather | Temp. | Barometer | Wind | Waves |
|---|---|---|---|---|---|
| _______ | _________ | _____ | _______ | __________ | _________ |
| _______ | _________ | _____ | _______ | __________ | _________ |
| _______ | _________ | _____ | _______ | __________ | _________ |
| _______ | _________ | _____ | _______ | __________ | _________ |
| _______ | _________ | _____ | _______ | __________ | _________ |
| _______ | _________ | _____ | _______ | __________ | _________ |

**Additional Notes:**________________________________________

________________________________________________________

## Using a Plan Effectively

The best way to use the plan outline is to *not* wait until just a few hours before you leave. Obviously, some information is better than no information, but consider getting weather data at least one day, and preferably two days, before your planned departure. That gives you a chance to verify the forecasts with observed conditions. If the forecasts are pretty close to what's observed, planning can proceed with more confidence than if the forecast and observed weather conditions are 180 degrees apart. Here's a suggested sequence for gathering information.

### *Two Days Before the Trip*

- Check the overall weather pattern.
- Check the projected weather for the next two days.

### *One Day Before the Trip*

- Check the current weather to evaluate the accuracy of the previous day's forecasts.
- Check the overall weather pattern.
- Check the projected weather for the next two days.
- If the possibility of high winds, snow, or thunderstorms is mentioned, plan on checking for updates every 6 to 8 hours. The lead time on such forecasts is short because of the rapid changes that can sometimes occur.

### *Day of the Trip*

- Check the current weather to evaluate the previous day's forecasts.
- Check the projected weather for the trip.
- Make a go/no-go decision based on the current forecasts, the track record of earlier forecasts, your personal experience and comfort level, the experience and comfort level of your crew, and the capabilities of your vessel.

Let's look at several examples to see how such information-gathering with a plan can be used effectively.

### *Marine Weather Checklist (Sample 1)*

**Planned Trip:** Depart from moorage on Lake Washington just east of Seattle at about 7:00 A.M. on Saturday, May 18, for a trip through the Hiram Chittenden locks into Puget Sound. Plan to pick up friends at J dock at Shilshole Bay Marina; then sail the central part of Puget Sound, stopping for dinner at Bainbridge Island; then make an evening sail for an overnight stay aboard in Port Madison.

**Date of Briefing:** Friday, May 17

| **Checklist** | **Data Sources** |
|---|---|
| X Synopsis | Weather Radio Frequencies KHB-60 @ 162.55 MHz |

| | | |
|---|---|---|
| X Warnings/Watches | Telephone Weather # NWS: | 526-6087 |
| X Zone Forecasts | KING-TV: | 448-3649 |
| X Marine Forecasts | Computer Data Source | WSI |
| X Offshore/Bar Conditions | Modem # | 747-1000 |
| X Extended Outlook | User ID | 12345 |
| X Local Observations | Password | HINDRY |
| X Barometer | | |
| X Notebook/Weather Logbook | | |

**Weather Synopsis:** A surface low is expected to pass across Vancouver Island during the early afternoon hours of Friday, May 17, the day before your planned trip. A cold front trailing from that low is expected to move through Puget Sound later that afternoon. The low itself is only moderately strong, with reported central pressure of 1008 millibars. High pressure will gradually build into western Washington overnight, with clearing forecast on Saturday, May 18, the day of your planned departure. *(Source: NOAA weather radio)*

**Warnings/Watches:** A small-craft advisory is in effect for the Strait of Juan de Fuca, the Washington coast, and from Camano Island north to Point Roberts. *(Source: NOAA weather radio)*

**Zone Forecast(s)**

| | Date: 5/17 | Date: 5/18 | Date: 5/19 |
|---|---|---|---|
| Sky/Weather | M. Cldy, Rain→Showers | A.M. Clouds→Clearing | Mostly sunny |
| Temp. (F) | Mid-50s | Mid- to upper 50s | Low 60s |
| Winds | SE 5–15 kt | SW 5–15 kt becoming N 5–15 | N 0–10 kt |

*(Source: NOAA weather radio)*

**Extended Outlook from Sunday, May 19, to Tuesday, May 21:** Mostly sunny and seasonably mild on Sunday and Monday, with daytime highs in the mid-60s and overnight lows in the upper 40s (Fahrenheit). Increasing clouds expected during the day Tuesday, with possible rain Tuesday night. Daytime highs Tuesday in the upper 50s.

**Marine Outlook:**

Location: Washington Coast
　　Wind/Waves/Swell: SE 15–25 kt/1–3'/SW 3–5'
Location: Strait of Juan de Fuca
　　Wind/Waves/Swell: E 10–20 kt; 15–25 kt entrances/—/—
Location: Camano Island N.
　　Wind/Waves/Swell: SE 5–20 kt/—/—
Location: Puget Sound
　　Wind/Waves/Swell: S 5–15 kt/—/—

*(Source: NOAA weather radio)*

**Weather Observations:** The observations Friday matched the forecast almost exactly. But early Saturday morning, you discovered interesting variations.

| Location | Sky/ Weather | Temp. (F) | Barom-eter(") | Wind (kt) | Waves (ft) |
|---|---|---|---|---|---|
| Quillayute | P. Cldy | 48 | 30.08 | W–NW 17 | 3 |
| Neah Bay | P. Cldy | 48 | 30.08 | W 13 | 2 |
| Port Angeles | P. Cldy | 47 | 30.03 | W 16 | — |
| Bellingham | P. Cldy | 47 | 30.00 | SW 15 | — |
| Whidbey | M. Cldy | 48 | 29.98 | W–NW 20 | — |
| Everett | M. Cldy | 45 | 29.98 | W–NW 15 | — |
| Seattle | M. Cldy | 46 | 29.97 | E 8 | — |
| Tacoma | M. Cldy | 46 | 29.97 | SW 7 | — |

*(Source: NOAA weather radio)*

**Additional Notes:** The sky looks very gray, with occasional drizzle. Some evidence of breaks in the clouds is seen to the south of the marina where you moor your boat on Lake Washington. Winds are light and variable.

**Analysis**: The forecast for Friday was correct, and the presence of some breaks in the clouds suggests the optimistic nature of the forecast for Saturday and Sunday may be correct. The clearing along the coast and the Strait of Juan de Fuca supports this, and typically low stratus present in the morning will clear, at least partially in the afternoon. The disturbing element is the quirky pattern of winds on Lake Washington, coupled with the onshore winds along the coast and strait. The fact that winds north of Seattle are from the northwest to west and that winds to the south are from the southwest is very suggestive of the Puget Sound Convergence Zone.

Remember that the convergence zone typically follows cold-front passage, with high pressure building in along the coast. A southwest-to-northwest flow along the coast splits around the Olympics, converging in central Puget Sound, usually between Everett and Tacoma. The NOAA weather radio wind reports just mentioned indicate exactly this pattern. The easterlies in Seattle are typical of the fluky winds in the middle of the convergence zone.

The collision of air from the north and south that gives the convergence zone its name results in rising air and often cloud formation or thicker clouds than might be found elsewhere. Also, showers typically occur, and occasionally thundershowers, even when not anticipated in the forecast. Gusty winds, ice pellets, and lightning strikes all offer good reasons to avoid an active convergence zone. Just as the actual convergence zone is marked by rising air and typically foul weather, the areas just to the north and south are usually marked by sinking air and considerably better weather.

Since weather conditions strongly point to the formation of the Puget Sound Convergence Zone and the possibility of showers or even thundershowers, it would be wise to choose an alternative sail, since Bainbridge Island and Port Madison are right in the midst of the convergence zone.

Depending upon the weather reports and what you see once you clear the locks and pick up your friends at Shilshole, you might play it safe and stick close to Shilshole, or head north to get beyond the convergence zone, perhaps along Whidbey Island or Camano Island.

### *Marine Weather Checklist (Sample 2)*

**Planned Trip:** You're under way on a trip you've waited years to make: a circumnavigation of Vancouver Island. Blessed with fair weather at the time of your departure from Victoria, you headed out through the Strait of Juan de Fuca and up the west coast. The weather has held, with relatively mild temperatures and fairly gentle seas, all the way from Victoria to Nootka. It's Wednesday, September 14, and you're planning to proceed from Nootka north tomorrow, knowing that storms can and do move through in September. You want to get around the north end of the island and into the more protected waters of Queen Charlotte Strait before such a storm does arrive.

**Date of Briefing:** September 14

| **Checklist** | **Data Sources** | |
|---|---|---|
| X Synopsis | Weather Radio Frequencies | WX2:162.40 MHz |
| X Warnings/Watches | Telephone Weather # | Port Hardy: 949-6559 |
| X Zone Forecasts | | Victoria: 363-6629 |
| X Marine Forecasts | Computer Data Source | Not Available |
| X Offshore/Bar Conditions | Modem # | ________ |
| X Extended Outlook | User ID | ________ |
| X Local Observations | Password | ________ |
| X Barometer | | |
| X Notebook/Weather Logbook | | |

**Weather Synopsis:** The high-pressure ridge that has been protecting the Pacific Northwest remains in place, but a small upper-air disturbance is flattening the top of the ridge in the Gulf of Alaska. That disturbance is forecast to drop down the east side of the ridge into southeastern Alaska tomorrow and then into coastal British Columbia, probably reaching the northern half of Vancouver Island in three days, on the 17th. (*Source: Environment Canada Weather Radio*)

**Warnings/Watches:** None (*Source: Environment Canada Weather Radio*)

**Zone Forecast(s)**

| | Date: 9/14 | Date: 9/15 | Date: 9/16 |
|---|---|---|---|
| Sky/Weather | Mostly sunny | Mostly sunny | Increasing clouds |
| Temp. (F) | Mid-60s | Mid-60s | Upper 50s/Low 60s |
| Winds | NW 5–15 kt | Light and variable | S to SE 10–20 kt |

(*Source: Environment Canada Weather Radio*)

**Extended Outlook from Friday, Sept. 16, to Sunday, Sept. 18:** Increasing clouds Friday, with rain developing Saturday, changing to showers on Sunday. Highs in the 50s.
(*Source: Environment Canada Weather Radio*)

**Marine Outlook:**
Location: Nootka
Wind/Waves/Swell: NW 5–15 kt/1–3'/NW 2–4'
Location: Solander Island
Wind/Waves/Swell: NW 5–15 kt/1–3'/NW 2–4'
Location: Sartine Island
Wind/Waves/Swell: NE 5–15 kt/1–2'/NW 2–4'
Location: Sandspit
Wind/Waves/Swell: SE 5–15 kt/1–3'/—
Location: Langara Island
Wind/Waves/Swell: SE 5–15 kt/1–3'/NW 2–4'
(*Source: Environment Canada Weather Radio*)

**Weather Observations**

| Location | Sky/ Weather | Temp. (F) | Barometer (mb) | Wind (kt) | Waves (ft) |
|---|---|---|---|---|---|
| Estevan Point | Clear | 62 | 1020 | NW 7 | 2 |
| Nootka | Clear | 60 | 1020 | NW 10 | 2 |
| Solander Island | Clear | 60 | 1018 | NW 8 | 1 |
| Sartine Island | Clear | 61 | 1017 | N 12 | 1 |
| Cape St. James | Clear | 59 | 1015 | NW 4 | 1 |
| Kindakun Rock | Mostly Clear | 62 | 1012 | N 3 | 1 |
| Langara Island | Mostly Clear | 59 | 1011 | Calm | 1 |

(*Source: Environment Canada Weather Radio*)

**Additional Notes:** Nothing exceptional other than fine weather. Shipboard weather facsimile charts indicate a weak surface low near Anchorage, with associated surface winds of 20 to 30 knots. That would be small-craft advisory range. Satellite image shows cloud cover extending from Fairbanks south through Kodiak and Anchorage over the Gulf of Alaska, with the eastern edge of clouds near Juneau. The upper-air chart shows that winds aloft at 18,000 feet range between 30 and 40 knots.
(*Source: Alden Weather Facsimile Receiver/NOAA weather charts*)

**Analysis:** The immediate outlook is excellent, with at least two days of really good sailing weather. Even the disturbance expected to arrive in three days, on Saturday, appears to be relatively mild. Important considerations in planning your course of action include likely cruising speed, including help or hindrance from currents, and the distance to be covered given that speed and the time you plan to spend under way each day.

Given the winds aloft, the disturbance will probably move about 20 nautical miles per hour, or close to 500 nautical miles per day. That would bring the disturbance to Sitka by this time tomorrow, on Thursday, and to

Cape St. James in 48 hours, by late Friday. That's probably fairly close to the time you'd be reaching Cape Scott and turning into Queen Charlotte Strait. Consider possible safe harbors if the system moves more rapidly and intensifies or if you cover less distance than anticipated.

You also need to consider the capabilities of your vessel, yourself, and your crew in handling the weather created by the disturbance. At present, that may not be of much concern or question. Do continue monitoring the weather reports and forecasts issued by Environment Canada and the surface and upper-air weather charts and satellite pictures coming over your facsimile receiver. If the surface disturbance begins to strengthen, or if you notice associated surface or upper-air winds beginning to strengthen, you have some important clues that the actual winds and sea states may be worse than presently forecast.

Carefully monitor your barometer for the rate of pressure change, which can indicate if the disturbance is moving more slowly or more rapidly than forecast. Also watch changes in sea state, sky cover, and winds, and compare these with previous reports and current forecasts. Also consider the effect of terrain on winds to determine whether wind channeling will increase velocities and sea states beyond those forecast.

### *Marine Weather Checklist (Sample 3)*

**Planned Trip:** Depart Friday, May 14, from Port Townsend, crossing the Strait of Juan de Fuca, with a possible stop at Roche Harbor, then continuing up Haro Strait to Bedwell Harbor to clear Canadian customs for a week of cruising in the Gulf Islands.

**Date of Briefing:** Thursday, May 13

| Checklist | Data Sources |
|---|---|
| X Synopsis | Weather Radio Frequencies KHB 60 @ 162.55 MHz, also KHB 162.475 (Victoria) |
| X Warnings/Watches | |
| X Zone Forecasts | Telephone Weather # NWS/Port Angeles: 457-6533 |
| X Marine Forecasts | Computer Data Source WSI |
| X Offshore/Bar Conditions | Modem # 747-1000 |
| X Extended Outlook | User ID 12345 |
| X Local Observations | Password MINNOW |
| X Barometer | |
| X Notebook/Weather Logbook | |

**Weather Synopsis:** It simply doesn't get much better than this. It's been very clear, thanks to a big high-pressure system dominating the Northwest. The high is centered over southcentral British Columbia, with a weak trough offshore. Surface winds have been offshore, producing clear skies, shallow morning ground fog, and unseasonably warm temperatures. Highs have been in the low to mid-70s Fahrenheit. The high is forecast to continue sliding to the southeast, into northern Idaho. There is no weather associated with the offshore trough, nor any lows of significance approaching. Another high sits just to the west of the offshore trough. (*Source: NOAA weather radio/personal observations*)

**Warnings/Watches:** None

**Zone Forecast(s)**

| | Date: 5/13 | Date: 5/14 | Date: 5/15 |
|---|---|---|---|
| Sky/Weather | Clear, sunny | Mostly sunny | A.M. clouds |
| Temp. (F) | Low to mid-70s | Around 70 | Low 60s |
| Winds | NE 5–20 kt | NE 5–20 kt | NW 5–15 kt |

(*Source: NOAA weather radio*)

**Extended Outlook from Saturday, May 15, to Monday, May 17:** Low morning clouds and fog, burning off by mid-afternoon early in the period, with daytime highs in the 60s and overnight lows in the upper 40s Fahrenheit. Then patchy morning clouds with mostly sunny skies, highs near 70. (*Source: NOAA weather radio*)

**Marine Outlook:**

Location: Washington coast
Wind/Waves/Swell: Light and variable/1–2'/NW 1–3'
Location: Strait/Admiralty Inlet
Wind/Waves/Swell: East 0–10 kt/—/—
Location: Camano Island to Point Roberts
Wind/Waves/Swell: NE 5–15 kt/—/—
Location: Puget Sound/Hood Canal
Wind/Waves/Swell: NE 5–20 kt/—/—
(*Source: NOAA weather radio*)

**Weather Observations**

| **Location** | **Sky/ Weather** | **Temp. (F)** | **Barometer (mb)** | **Wind (kt)** | **Waves (ft)** |
|---|---|---|---|---|---|
| Port Angeles | Clear | 62 | 1008 | E 3 | — |
| Neah Bay | Clear | 60 | 1008 | Calm | 1 |
| Victoria | Clear | 65 | 1009 | SE 2 | — |
| Quillayute | Clear | 63 | 1008 | Calm | 1 |
| Long Beach | Clear | 61 | 1009 | E 3 | 1 |
| Astoria | Clear | 60 | 1010 | E 5 | 1 |
| North Bend | Fog | 54 | 1011 | SW 15 | 2 |
| Seattle | Clear | 69 | 1008 | NE 3 | — |
| Everett | Clear | 68 | 1008 | E–NE 4 | — |

(*Sources: NOAA weather radio/WSI*)

**Additional Notes:** Television forecasters report warm temperatures east of the Cascades, with Yakima at 85 degrees Fahrenheit, Wenatchee at 81, and the Tri-Cities at 87. You also notice that satellite images indicate a fingerlike extension of clouds along the south Oregon coast.

**Analysis:** On the surface, the weather forecasts point to excellent weather for the crossing. Winds are forecast to be light, with seas as close to calm as you can get and sunny, mild weather. The Saturday forecast does call for low morning clouds and a shift to onshore winds, out of the southwest, but by that time you'll be up in the Canadian Gulf Islands and well past the Strait of Juan de Fuca. But there are some warning signs that perhaps you need to keep an eye on.

First, the unseasonably warm temperatures on both sides of the Cascades often are ended by an abrupt shift to an onshore flow, cooler temperatures, low stratus or fog, and gusty winds. This "warm wave" has been under way for several days. Second, the fingerlike extension of clouds along the south Oregon Coast quite literally points to such an onshore push as the surge of cooler ocean air moves up the coast.

The third warning sign is the air pressure shift and the difference between some key locations. The pressure is higher at the coastal stations of North Bend, Oregon, and Astoria, Oregon, than at Seattle, with a difference of 2 millibars between Astoria and Seattle and 3 millibars between North Bend and Seattle. This is a fairly reliable signal that a push that will extend up the rest of the Oregon and Washington coast and into the interior is under way. When cooler ocean air surges through the Columbia River valley and the Strait of Juan de Fuca, there's an abrupt wind shift to westerlies. If this occurs prior to or during your departure, the crossing would be, at the very least, unpleasant.

Right now, the pressure differences are marginal for such an onshore push. So plan to get up early in order to see whether the pressure differences have become greater, whether the shift to westerlies has advanced northward along the coast, and whether satellite images show the finger of clouds has surged northward along the coast. If the action is still limited to the Oregon coast, a crossing of the strait will probably be safe if it's conducted promptly during the morning. Waiting until the afternoon could be stretching your luck. If the stratus and onshore winds have moved up along the Washington coast by the morning, odds are you'll get caught by the push. You can evaluate your ability to handle the likely winds by checking winds at that time along the Columbia River. The winds in the strait will probably be similar.

The preceding examples show how personal observations, paired with official weather reports and forecasts and a measure of weather know-how, can guide you in making good decisions. This never precludes the possibility of surprises, however. That's the reason for the final two chapters, which offer some field forecasting tips for the waters of the Pacific Northwest, followed by a summary of localized weather patterns.

CHAPTER 8

# Field Forecasting Guidelines

*You don't need a weatherman*
*To know which way the wind blows*

Bob Dylan
"Subterranean Homesick Blues"

Weather-related accidents on the water rarely occur without warning. Occasionally the clues are subtle, at other times broad as daylight. Caution can't end with the decision to cast off or weigh anchor; anyone who has spent time on the water knows that weather forecasts can and do go wrong. Skippers also have to be ready to do some of their own field forecasting based on a few tested guidelines.

Even the best professional forecasters have their off days. Many of the busted forecasts aren't due to human error but are the result of limited computer ability to digest a mountain of weather observations, run them through complicated equations, and develop projections. The computer forecast models now in use are immeasurably better than those used just five years ago, but they are gluttons for computer memory. As impressive as new supercomputers are, they can't fit in all the data necessary to accurately assess the evolution of a weather pattern in the complex terrain of the Pacific Northwest. Previous chapters have shown the wide differences in weather over short distances in the Northwest caused by variations in terrain. As of this writing, the coastline was not yet accurately positioned, and the Olympics and Cascades weren't factored into the models; the computer only pictured a gradual upslope from the West Coast to the Rockies. Connect an airplane autopilot to the computer model and it would confidently fly due east from the coast directly into the mountains.

Another problem faced by the computer models is that they are a little like fishnets, with observing stations forming the intersections of strands. Just as fish smaller than the mesh can swim through the net, so can weather disturbances smaller than the "mesh" of the computer weather models slip through unnoticed. The opportunity for such slips is great, given the few observations available from over the Pacific.

Forecasters experienced in a region's weather patterns can sometimes correct errors in the computer projections, but not always. There is much about the atmosphere still to be learned; meteorology is a young science.

The wide variety in Northwest weather and the possibility of occasional incomplete or inaccurate forecasts means that skippers must understand weather and be able to do some field forecasting on their own. This chapter won't break new ground so much as it will organize the principles previously discussed into flowcharts to guide you in your onboard decision-making. This is the section you'll want to refer to at the helm or in the cabin. It is no substitute for current forecasts and observations; and like a good anchor, the field forecasting guidelines don't replace sound judgment and decision-making, they enhance it.

## Indicators of Approaching Weather Systems

There are four major indicators of an approaching storm: changes in cloud cover, in wind speed, in wind direction, and in air pressure. No single indicator is infallible; each should be examined.

*Cloud Cover Clues*

| *IF* | *THEN* | *CHECK FOR* |
|---|---|---|
| High cirrus clouds forming loose halo around sun/moon | Precipitation possible within 24 to 48 hours | Lowering, thickening clouds |
| High cirrus clouds forming tight corona around sun/moon | Precipitation likely within 24 hours | Lowering, thickening clouds |
| "Cap" or lenticular clouds forming over peaks | Precipitation possible within 24 to 48 hours | Lowering, thickening clouds |
| Thickening, lowering, layered/flat clouds | Warm/occluded front likely approaching within 12 to 24 hours | Wind shifts, pressure drops |
| Breaks in cloud cover closing up | Cold front likely within 12 hours | Wind shifts, pressure drops |

*Clues from Winds Aloft*

| *IF* | *THEN* | *EXPECT* |
|---|---|---|
| Cirrus clouds moving from SSW to NNE | Almost continuous moderate to heavy precipitation for more than 24 hours | 1 to 2 inches of rain inland, more than 2 inches along the coasts, and strong southerly winds |
| Cirrus clouds moving from WSW to W | Moderate precipitation is likely, with breaks between systems | Possible thunder-storms, particularly along mountains |
| Cirrus clouds moving from WNW to NW | Fast-moving cold front is likely, with moderate precipitation | Thunderstorms, ice pellets, rain, and/or snow showers likely within 48 hours of cold-front passage |
| Clouds moving from N and freezing level at or near sea level | Arctic air mass moving into area | Gusty, cold winds out of east–west rivers draining from Cascades/Coast Mountains; velocities of 60 knots or more possible, with snow or freezing rain |
| Clouds moving from NE | Sunny weather likely, with above-normal temperatures in summer, below-normal in winter | Possible temperature inversion in two or more days, with lowered visibility in basins |

*Clues from Surface Winds*

| *IF* | *THEN* |
|---|---|
| Surface winds veer to N or NE | Expect clearing |
| Surface winds veer to E or SE and increase | Watch for increasing clouds/precipitation |
| Surface winds veer to SW to NW | Frontal passage has occurred; cooling possible, with occa-sional showers likely |
| Surface winds back to E or SE and increase | Low approaching; watch for increasing clouds, wind; expect precipitation |

### *Clues from Pressure Changes*

Remember that a barometer can give excellent warning of an approaching weather system. As with all guidelines, the following summary is no substitute for obtaining a complete forecast, but it can be a useful aid.

Changes in pressure create changes in wind and are often related to approaching fronts that may bring precipitation. Notice that pressure indications are given both in inches of mercury and in millibars, and that action is specified for pressure *changes,* not simply pressure *decreases.* A rapidly building high indicated by increasing pressure can generate strong winds, too.

| *3-hour Pressure Change* | *Advised Action* |
|---|---|
| .02–.04" (.6–1.2 mb) | None. Continue to monitor. |
| .04–.06" (1.2–1.8 mb) | Clouds lowering/thickening? If so, begin checking pressure changes hourly. |
| .06–.08" (1.8–2.4 mb) | Small-craft advisory conditions likely. Consider finding protected harbor. Check pressure changes hourly. |
| More than .08" (more than 2.4 mb) | Gale-force winds likely. Find protected harbor and check pressure changes hourly. |

## Thunderstorm and Lightning Safety Guidelines

The best safety advice for thunderstorms is simple: Avoid them. Here are some guidelines on what to do if caught near one.

### *If Thunderstorms Are Forecast*

- First watch small cumulus for strong upward growth.
- Keep track of weather reports hourly.
- Listen for strong static interference on AM radio broadcasts, which may come from lightning.
- Consider altering course to keep you near safe anchorages.

### *If You Spot Thunderstorms or Hear Thunder*

- Seek safe anchorage immediately if possible.
- If safe anchorage is not possible, try to seek safe shelter inside your craft.
- Prepare for gusty winds in rough water.
- Track the movement of the thunderstorms.

To track the movement of a thunderstorm, start timing when you see lightning. Stop timing when you hear the thunderclap. Divide the number of seconds by 5 for the distance from the storm in miles.

| *IF* | *THEN* |
|---|---|
| Time interval increasing | Storm moving away |
| Time interval decreasing | Storm approaching |

## Fog Clues

| *IF* | *AND IF* | *THEN* |
|---|---|---|
| Ground wet, pressure increasing after frontal passage | Sky clearing overnight, winds 5 knots or less | Radiation fog likely by or shortly after sunrise; expect burnoff by noon unless inversion is present |
| Above-average temperatures over region during late spring or summer months | Fog/stratus advancing northward along Oregon/Washington coasts | Advection fog likely with possible drizzle, and strong wind shift in east–west channels to westerly winds prior to fog arrival; check specific area sailing directions |
| Thickening, lowering clouds, falling air pressure | Rain falling into cold air, especially in fjords/basins | Warm frontal fog likely; won't clear until after warm-front passage |
| Clearing after frontal passage and considerable cooling, especially in autumn months | Water temperatures still relatively warmer than air | Steam fog likely; will typically clear by mid-morning |

CHAPTER 9

# Sailing Directions: A Skipper's Reference to Local Weather Phenomena

*We had scarcely finished our examination, when the wind became excessively boisterous from the southward, attended with heavy squalls and torrents of rain which continuing until noon the following day, occasioned a very unpleasant detention.*

Log of Captain George Vancouver, June 1792,
Howe Sound

Weather has been surprising even the best of mariners for as long as they've been casting off for destinations both exotic and familiar. Perhaps the most valued possession of the master of any vessel in years past was a small notebook filled with guidance gleaned from experienced salts, describing the local hazards and suggested strategies along a route. Such lore was the greatest possible gift from a navigator who had explored new waters to those who would follow in his wake.

This final chapter is written in that spirit: it's a specific listing of the most common weather hazards and strategies for the various waters of the Pacific Northwest.

## Sailing Directions for Specific Cruising Zones

One of the beauties and frustrations of the Pacific Northwest is that weather can change considerably just around a point or inside a cove. It's virtually impossible to issue a forecast encompassing all of the potential variations in an area. That's the purpose of the sailing directions in this chapter. They offer weather guidance for specific cruising areas. Select the appropriate area, read the climate summary as a pre-trip briefing, and then check the planning map provided, looking for the symbols depicting various weather challenges along or near your intended course. Read the description of the hazard, and the key indicators, and consider the suggested strategy.

Each planning map lists the locations and names of weather stations that report observations on either NOAA or Canadian weather radio. Along with the general forecasts issued, those observations can help confirm the possibility of such localized weather phenomena. The guidelines that follow are not intended to replace forecasts; they're meant to augment such forecasts on the water. There's no substitute for an updated forecast from a professional meteorologist; obtaining one should precede every sail, cruise, or paddle. The thoughtful skipper will carry a weather radio for updates.

# The Columbia River

## From the Gorge to Astoria

### *Climate Summary*

Dramatic variations in weather can be found over the short distance from the mouth of the Columbia River (near Astoria, Oregon) to the Columbia River Gorge. A maritime climate exists near the mouth, gradually changing to a continental climate upriver. Weather hazards include strong winds associated with approaching lows and funneling, rough water caused by wind/current/wave interactions, and fog.

The Columbia River west of the gorge has a winter rainfall climate. Portland receives 88 percent of its rainfall during the months of October through May; only 3 percent falls during July and August. Snow is infrequent, most falling from December through February.

The east–west orientation of the river leads to an interesting clash of cold air from eastern Oregon and warm, moist air from the Pacific. As a low approaches the coast, subfreezing air is drawn from the east. The warm, moist ocean air rides up and over the cold air, typically producing extensive freezing rain between the gorge and Portland, coating everything in a shiny, slippery veneer of ice. Eventually the cold air warms and the freezing rain changes to rain. This pattern of ice followed by warming resulted in the name "silver thaw."

Prevailing winds are from the east during winter and from the west during summer. At The Dalles, peak velocity may vary during a single day from 6 to 42 knots. From April through August, winds reach at least 17 knots about two to three days each week. That's why the area is a haven for windsurfers.

Those prevailing summer westerlies also bring in fog, which becomes a nuisance near the mouth of the Columbia during the summer and moves upriver by September. Visibility drops below one-half mile in and near Portland roughly four to eight days a month during the late summer to mid-autumn months. At times, the shift from easterly winds to westerlies can be dramatic, carrying in a wall of fog or low stratus. This usually follows a run of 90-degree-plus temperatures east of the Cascades. The hot air in that region rises, creating what's called a thermal low. The shift is simply the movement of cooler, higher pressure air from the coast to replace that rising air. The sudden change is just another example of the need to be familiar with local weather patterns and to keep current on changes in weather conditions and forecasts.

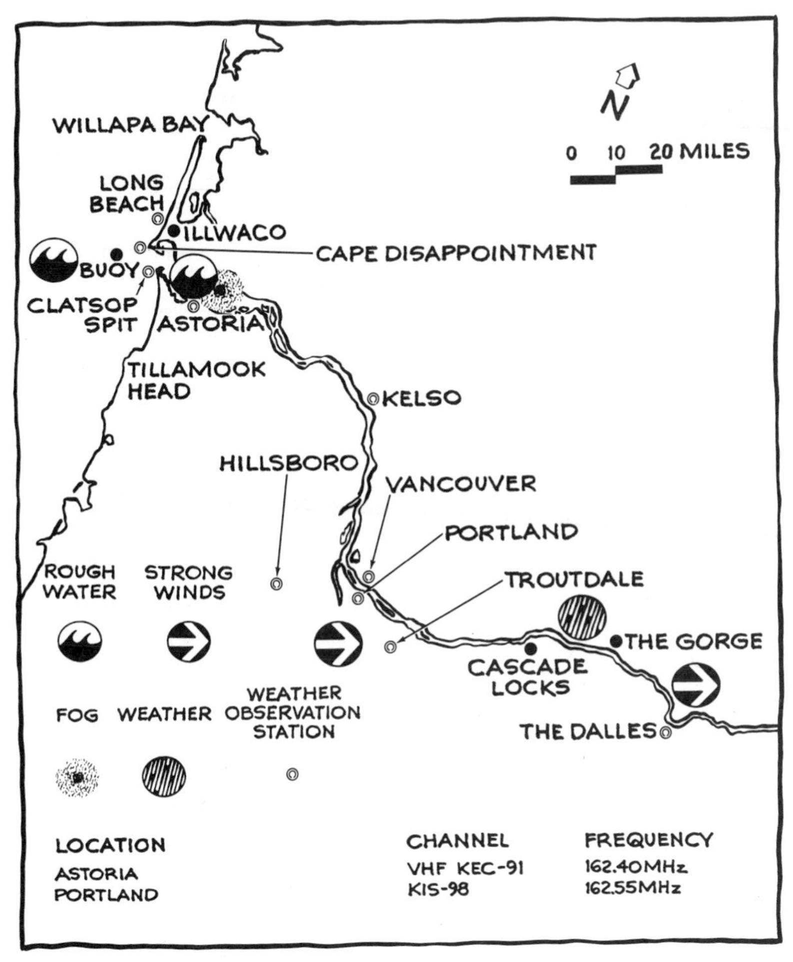


*The Columbia River, from the Gorge to Astoria*

## Sailing Directions:

*The Columbia River, from the Gorge to Astoria*

| *Weather Hazard* | *Location* | *Specific Caution* |
|---|---|---|
| Strong Winds | Gorge | Strong winds are the result of a strong pressure gradient. The following guidelines are approximations of the winds typically produced by various pressure gradients (differences) between Portland and The Dalles.<br>5 mb→20- to 30-knot winds;<br>10 mb→30- to 40-knot winds;<br>15 mb→40 knots or more. |
| Strong Winds | Portland/ Vancouver, Wash. | Strong winds are again the result of strong pressure gradients through the Columbia River. On average, a 10 mb difference between Portland and The Dalles suggests near gale-force or gale-force winds. |
| Strong Winds | Portland/ Vancouver, Wash. | During winter, a low swinging inland north of Vancouver can generate strong southerly winds, both because of the strength of the low and because of funneling between the Coast and Cascade ranges. Be particularly vigilant when approaching lows have a central pressure of 975 mb or less; storm-force winds are possible. |
| Rough Water | Columbia River Bar | The bar is subject to sudden unpredictable changes in current, which rapidly produce breakers. These breakers can extend as far as Buoy 20, north of Clatsop Spit. Particularly dangerous is the combination of an ebb tide and westerly winds/ waves/swell. It's best to wait for slack water, or at least a flood tide. |

## Sailing Directions:

*The Columbia River, from the Gorge to Astoria*

| *Weather Hazard* | *Location* | *Specific Caution* |
|---|---|---|
| Rough Water/Fog | Columbia River Bar inland | The combination of fog and currents make it all too easy to run aground. Experienced skippers recommend remaining outside the 30-fathom curve until reaching the Columbia River approach lighted horn buoy. |
| Strong Winds/Fog | Astoria east to Portland | Hot weather east of the Coast Range and in the Willamette Valley generates a strong pressure difference through the Columbia River Valley. Cooler ocean air surges inland, shifting winds from east to west and dropping visibilities in the onshore push. This is most common from May through September. Watch for the northward advance of fog and stratus along the Oregon coast and a wind shift from northerly or northeasterly to southerly or southwesterly. |
| Freezing Rain | Portland/ Vancouver, Wash., to the gorge | Freezing rain is likely when cold air moves through the gorge from eastern Oregon during the winter. Watch for subfreezing temperatures in eastern Oregon, a shift to strong easterlies along the gorge, and an approaching low producing rain along the coast. |

# The Washington Coast

## From the Columbia River to the Strait of Juan de Fuca

### *Climate Summary*

The weather of the Washington coast is every bit as rugged as the sheer cliffs and headlands that mark land's end. Lows that strengthen in the Gulf of Alaska make frequent direct hits. Gale-force winds are frequent during the winter rainy season, and storm-force winds are not uncommon. The wet season runs from September through March, with downpours giving way to drizzle and fog during the "dry" season.

Sunshine is no stranger to the coast, however, and hot weather with temperatures in the 80s or even 90s Fahrenheit occurs most frequently from mid-July through August. In fact, when high pressure builds inland, producing northeasterly, offshore winds, some of the warmest temperatures in western Washington are found along the northern Washington coast.

Southeasterly winds prevail from late September through March; north to northwesterly winds prevail during the late spring and summer months of April through September. Heavy, shocklike seas are common during the wet season near the mouth of the Strait of Juan de Fuca, when prevailing southeasterly winds meet southwesterly ocean swell. Overall, December and January are the windiest months, with the largest waves, along the southern Washington coast. That peak occurs a little later, in January and February, along the northern Washington coast. The Quinault River tends to serve as the dividing line between north and south for this purpose.

Thick radiation fog hugs the shore and much of Grays Harbor and Willapa Bay during September and October, with only partial afternoon thinning. Such fog typically follows the onset of light wind and clearing after rainy weather and is usually confined close to land.

Westerly winds bring the sea fog inland along much of the coast during both summer and autumn. Sea fog is more resistant to burning off and can't be escaped by heading a little farther out from land. The fog intensifies or increases its coverage during prolonged periods of northwesterly winds. Such winds push away the warmer surface water, permitting upwelling of cold subsurface waters, which rapidly cause condensation of the warmer air above. Several days of northwesterlies are usually needed for this to occur.

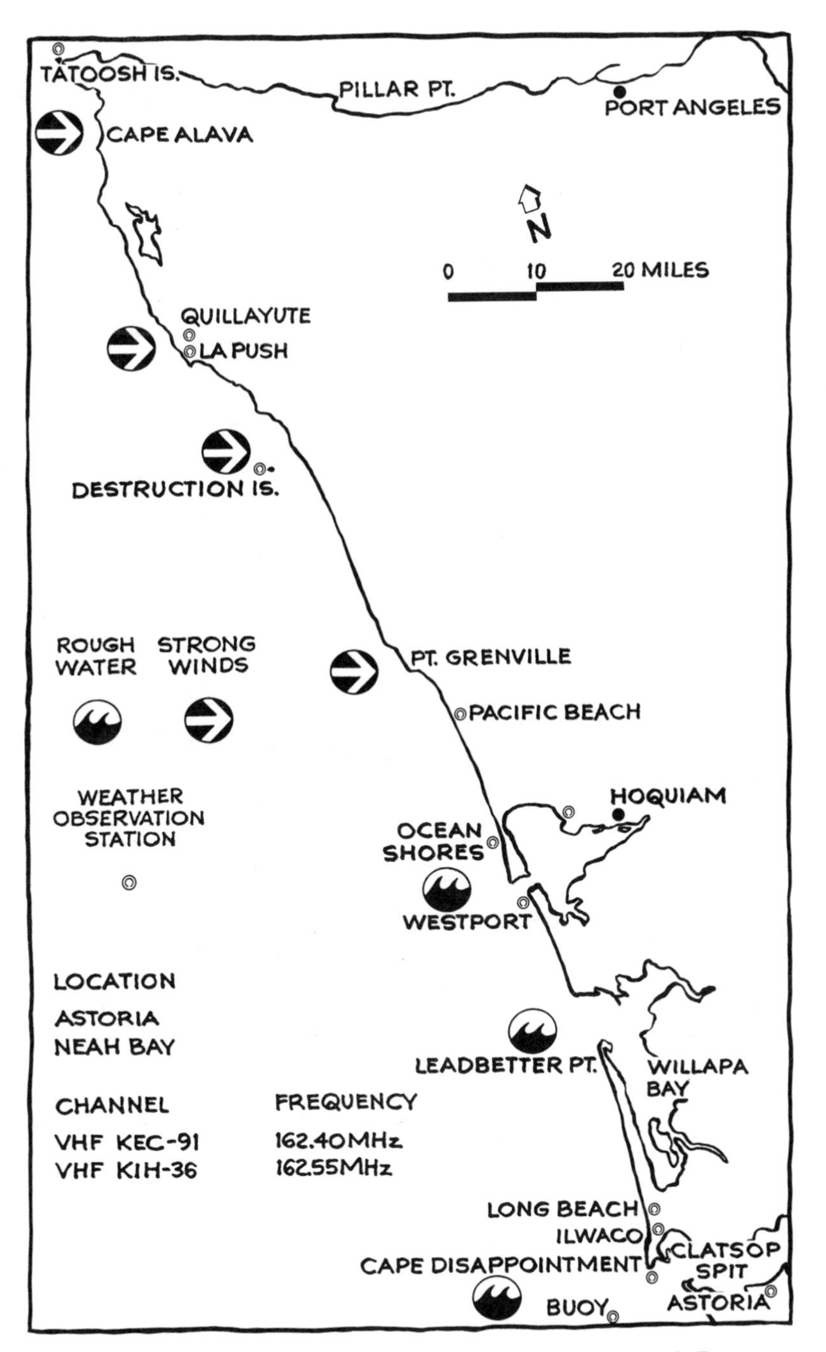


*The Washington coast, from the Columbia River to the Strait of Juan de Fuca*

## Sailing Directions:

*The Washington Coast*

| *Weather Hazard* | *Location* | *Specific Caution* |
|---|---|---|
| Strong Winds | Entire coast | Gale-force winds are commonly found along the coast when the pressure gradient (difference) between Astoria and Quillayute reaches or exceeds 4 mb. |
| Strong Winds | Entire coast | Gale-force winds commonly occur along the coast after frontal passage is followed by a strong high with central pressure of 30.40 inches (1030 mb) or greater. |
| Wind Funneling | Entire coast | Gale-force winds may occur along the coast even when ships or buoys report winds associated with an approaching low are lighter. This acceleration is due to air converging between the coast and the low. As the low approaches within 100 to 200 miles, wind speeds typically increase by 25 percent. This effect is most pronounced 1 to 2 miles offshore. |
| Wind Funneling | Tatoosh, La Push, and Destruction Island | Locally strong winds double the prevailing velocity are most common when prevailing winds are generally from the east. |
| Lee Winds | Point Grenville, Cape Alava, and Cape Flattery | Gusty lee winds and choppy seas are common when prevail-ing winds are from the southeast or north. |

## Sailing Directions:

*The Washington Coast*

| *Weather Hazard* | *Location* | *Specific Caution* |
|---|---|---|
| Rough Seas | Entrances to Grays Harbor, Willapa Bay, and the Columbia and other rivers | Rough bar conditions are common when easterly winds blow against swell or waves. Such rough seas typically lag behind peak winds by 1 to 6 hours. |
| Rough Seas | Entrances to Grays Harbor, Willapa Bay, and the Columbia River | Rough bar conditions are common when ebbtide currents run against waves. Such conditions are most likely to be dangerous when wave periods are less than 8 seconds and tidal currents equal or exceed 4 knots. |

# The Strait of Juan de Fuca and Admiralty Inlet

*Climate Summary*

Wind and fog are the two biggest hazards facing mariners in the Strait of Juan de Fuca. Winds reach their greatest strength from October through March. Gale-force winds are reported an average of four to six days per month during that period near the west and east entrances, and two to four days per month in the central part of the strait. The typical interval between winter storms ranges from one to five days at most.

As a low approaches, winds generally are not particularly strong within the strait but can easily reach gale force near the entrances, particularly the east entrance and Admiralty Inlet. The south shore of the strait tends to be protected from such southeasterly gales. Port Angeles offers good shelter during such blows. Once the low or associated front moves inland, the flow reverses to westerly throughout the strait, with the exception of the east entrance, where southwesterly winds are more frequent. The strongest winds in this area typically occur several hours after frontal passage.

During the summer months, prevailing winds are from the southwest to northwest, with calm conditions in the early morning and sea breezes reaching up to 30 knots by mid-afternoon. Forty-knot sea breezes aren't uncommon. For that reason, crossings are best planned for the early morning hours during fair weather when sea breezes are likely to be the predominant wind pattern. The sea breeze usually diminishes by early evening.

Sea fog is at its worst from July through October. The sea breeze typically carries the fog into the strait. It's an unpleasant combination: visibility less than half a mile and winds of 25 to 30 knots. Sea fog tends to penetrate farther east along the south shore than the north shore of the strait. It's more likely to reach Port Townsend than Victoria, and skies are generally clear north of Race Rocks. After prolonged hot weather in the interior, the shift from easterly to westerly winds can be breathtaking, rapidly followed by fog and drizzle.

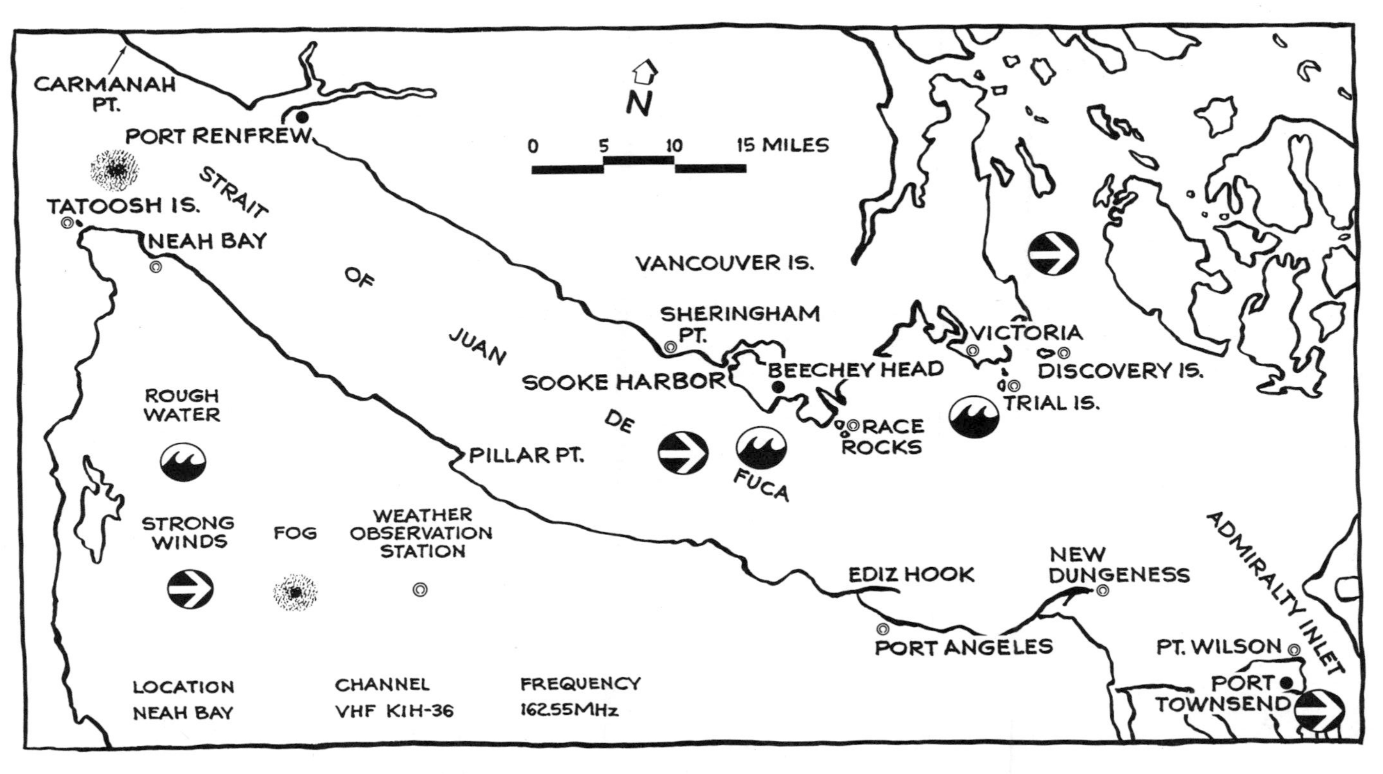


*The Strait of Juan de Fuca and Admiralty Inlet*

## Sailing Directions:

*The Strait of Juan de Fuca and Admiralty Inlet*

| *Weather Hazard* | *Location* | *Specific Caution* |
|---|---|---|
| Strong Winds | Entire strait/ Admiralty Inlet | Prevailing easterly or westerly winds will be channeled and funneled into the strait, with velocities as much as double the prevailing speeds. Gale-force winds are likely when the Quillayute/Bellingham pressure gradient reaches or exceeds 4 mb. Highest winds with westerlies are found near the east entrance/Admiralty Inlet. Highest winds with easterlies are found near the west entrance. |
| Strong Winds | East entrance/ Admiralty Inlet | Hot weather with offshore, easterly winds is typically followed by an abrupt shift to strong westerlies and cooler, foggy weather. If the Astoria/ Seattle pressure gradient exceeds 2 mb (higher pressure at Astoria) and fog/stratus is moving northward along the coast, anticipate the onset of an onshore push with 30- to 40-knot winds. |
| Strong Winds | Entire strait | Sea breeze develops most frequently from late spring to early autumn when skies are at least partially clear and prevailing morning winds are light. Velocities can reach 30 to 40 knots. Crossings are best planned for early morning hours. |
| Rough Seas | Race Rocks | Tidal currents can reach speeds of up to 6 knots. If wind direction opposes the current, dangerous steep seas can occur. |

## Sailing Directions:

*The Strait of Juan de Fuca and Admiralty Inlet*

| *Weather Hazard* | *Location* | *Specific Caution* |
|---|---|---|
| Rough Seas | Discovery Island | Strong rips tend to be amplified by easterly winds. |
| Rough Seas | West entrance/ Beechy Head | Ebb tides paired with westerly swell or wind-generated waves tend to produce short, choppy seas. Wait for flood tide. |
| Fog | West strait | Sea fog is especially common during late summer and autumn months. It's typically pulled farther east along the Washington shore than along Vancouver Island. |

# Puget Sound and Hood Canal

## *Climate Summary*

Although hardly dry, Puget Sound receives significantly less rainfall than coastal areas thanks to the sheltering influence of the Olympic Mountains. Because the Olympics are upwind of Puget Sound, the range tends to cast a rain shadow over the central and northern part of the sound. The wet season runs from October through April; approximately 82 percent of the yearly rainfall occurs during that period. The driest weeks of the year range from mid-July to mid-August. Snow is most likely from December through February.

Prevailing winds are generally southeasterly to southwesterly from September through April. Southerly winds typically produce the strongest velocities, although an arctic blast can generate high wind speeds from the north. During the late spring and summer months of May through August, prevailing winds are northwesterly to northerly. The warmest summer weather tends to occur when winds are from the north to northeast, a direction that also produces the coldest winter weather. When high pressure dominates the region and sea breezes occur, wind speeds are typically light and variable at night, rising to between 8 and 15 knots during the afternoon and early evening.

Fog reduces visibility significantly in the Puget Sound area on twenty-five to forty days of the year, with the most frequent occurrence of fog south of Tacoma or just north of the sound in Admiralty Inlet. There are two prime periods of foggy weather: during the months of September and October and then again during January and February. During these two periods, temperature inversions trap cold air, smog, and moisture in a thick, gray soup.

## Sailing Directions:

*Puget Sound and Hood Canal*

| *Weather Hazard* | *Location* | *Specific Caution* |
|---|---|---|
| Strong Winds | Entire sound | Gale-force winds are most likely in the sound when prevailing winds are from the north or south and the pressure gradient between Bellingham and Olympia is 4 mb or greater. |
| Strong Winds | Entire sound | The sea breeze builds from the north end of Puget Sound toward the south, first along the west side of the sound, then the east side. It tends to persist latest along the east side of the sound in the late afternoon to early-evening hours. |
| Strong Winds | Entire sound | A surge of moist ocean air and an abrupt shift from northeasterly to southwesterly winds tends to move southward into the sound after several conditions are met: unseasonably warm weather in the Puget Sound region and east of the Cascades; the northward progression of fog/stratus along the Oregon and Washington coasts; and a pressure gradient (difference) of 2 mb or more between Astoria and SeaTac (Seattle), with the higher pressure reading at Astoria. The gusty winds typically move into Puget Sound in the afternoon to early-evening hours, with low stratus or fog forming by sunrise the next day. |
| Strong Winds | East shoreline between Elliott Bay and Commencement Bay | Small-craft advisory or gale-force winds can occur when high pressure is centered east of the Cascades and the pressure gradient between Wenatchee and SeaTac (Seattle) is 10 mb or greater. |

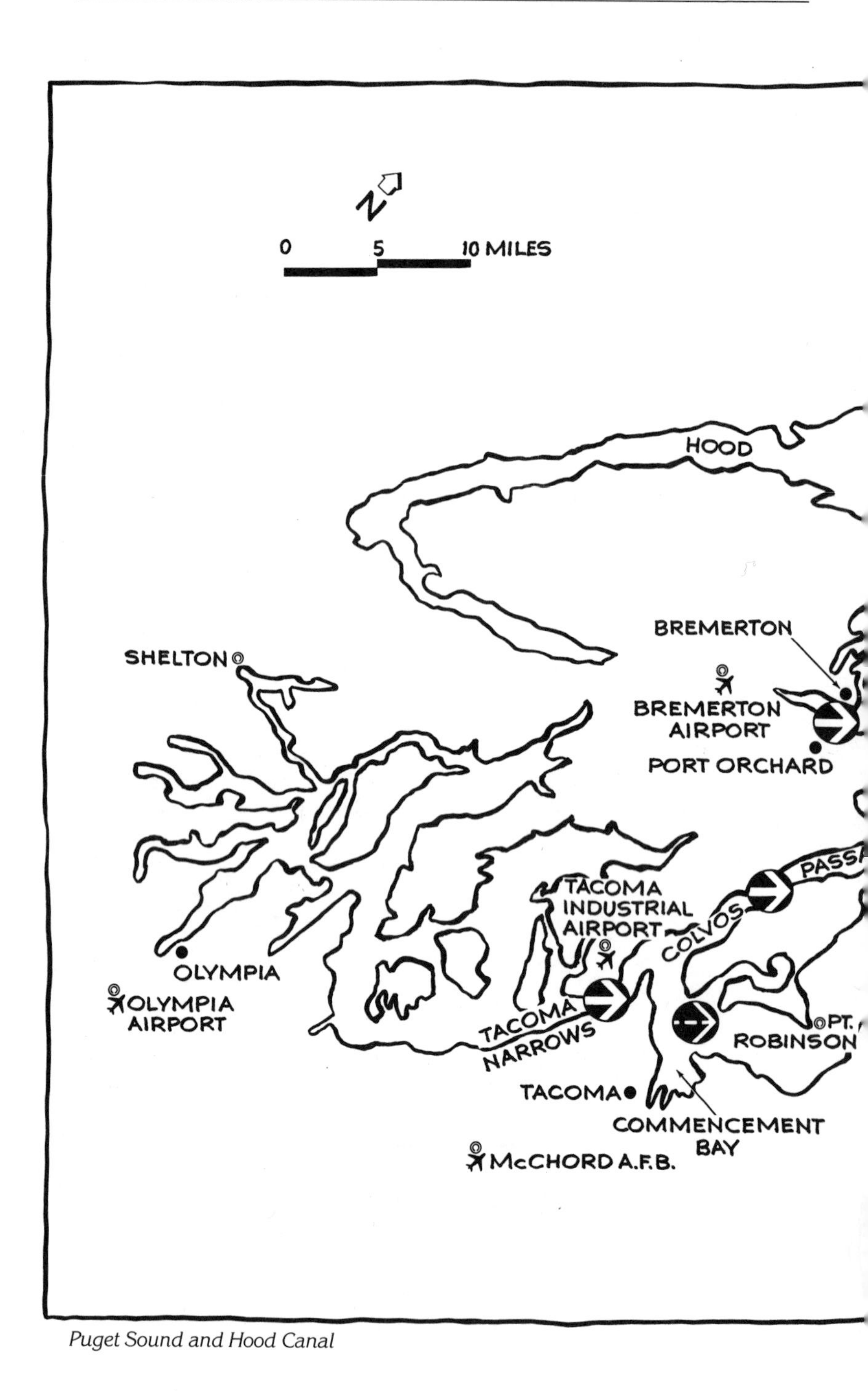


*Puget Sound and Hood Canal*

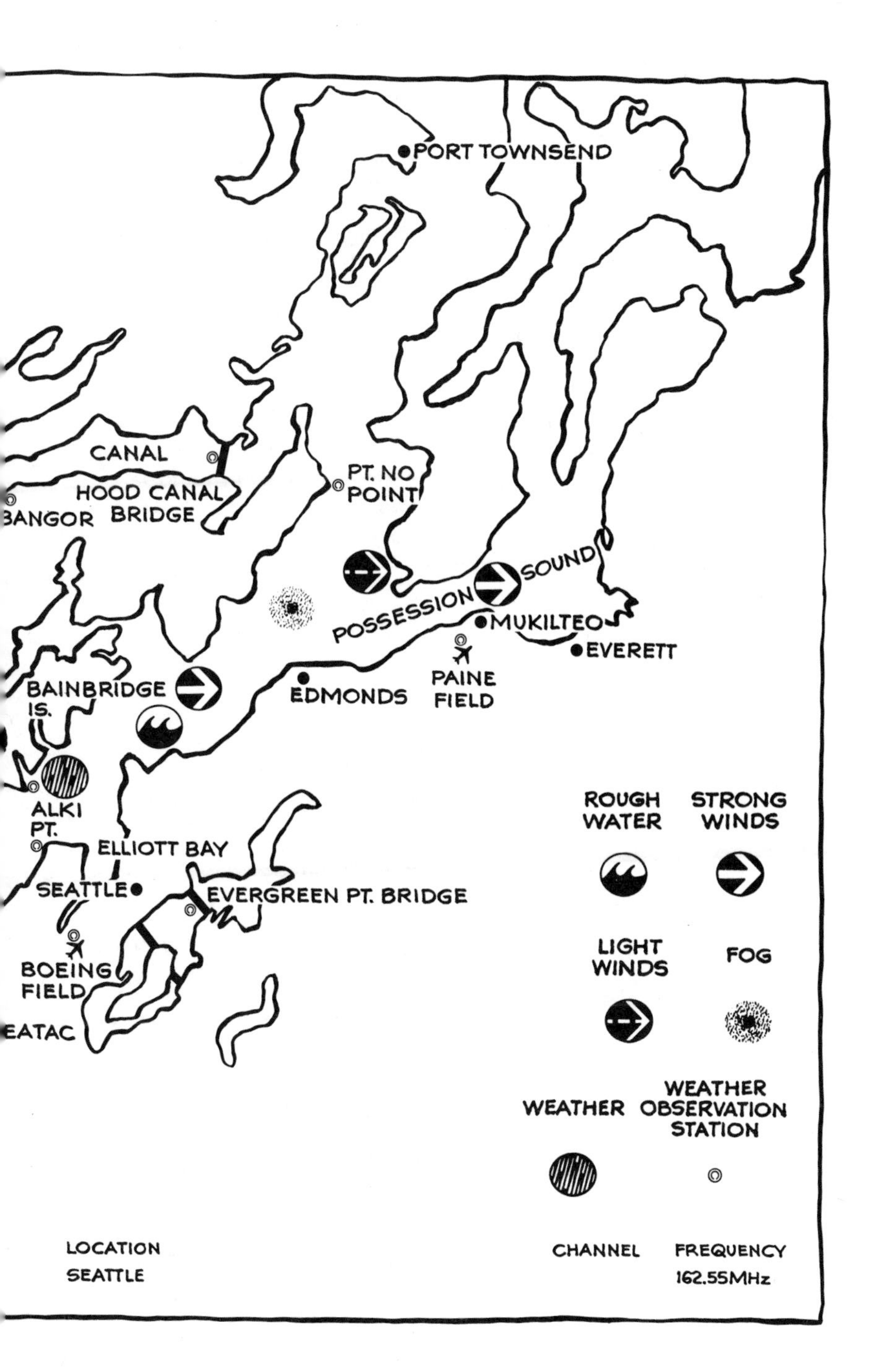
PORT TOWNSEND
CANAL
HOOD CANAL
BANGOR
BRIDGE
PT. NO POINT
SOUND
POSSESSION
MUKILTEO
EVERETT
PAINE FIELD
EDMONDS
BAINBRIDGE IS.
ALKI PT.
ELLIOTT BAY
SEATTLE
EVERGREEN PT. BRIDGE
BOEING FIELD
EATAC
ROUGH WATER
STRONG WINDS
LIGHT WINDS
FOG
WEATHER
WEATHER OBSERVATION STATION
LOCATION
SEATTLE
CHANNEL
FREQUENCY
162.55MHz

## Sailing Directions:

*Puget Sound and Hood Canal*

| *Weather Hazard* | *Location* | *Specific Caution* |
|---|---|---|
| Strong Winds | Colvos Passage, Tacoma Narrows, Port Orchard, and Possession Sound | Locally strong winds double the prevailing velocities are most likely if prevailing winds are from the north or south. |
| Strong Winds | East shoreline between Mukilteo and Edmonds, although can be found elsewhere | Gravity winds tend to provide fair sailing conditions after sunset. Such downslope winds are most likely when skies are partly cloudy to clear and sea breezes are the main wind regime during the day. |
| Wind Shadow | South end of islands/peninsulas, especially Whidbey and Vashon | Local terrain tends to provide a wind shadow, blocking northerly winds the most. |
| Strong Winds/ Rough Water/ Precipitation | Between Everett and Tacoma | The Puget Sound Convergence Zone produces locally variable winds, lowered ceiling, and increased precipitation (and occasionally thundershowers). This usually follows cold-front passage and southwesterly to west-northwesterly winds along the coast. Winds tend to be northerly to northwesterly to the north of the zone and southerly to southwesterly to the south of the zone. Considerably better weather is found just to the north and south of the zone. |
| Fog | Entire sound | Radiation fog generated by clearing and cooling after precipitation is most frequent and most dense south of Tacoma during September, October, January, and February. Temperature inversions may develop, causing the fog to persist for days, with limited afternoon clearing. |

# The San Juan Islands and Gulf Islands

## From Victoria East to Anacortes, North to Vancouver, and West to Nanaimo

### *Climate Summary*

The San Juan Islands and the Gulf Islands enjoy a much sunnier and drier climate than the rest of the region covered in this book. Both island groups are in the lee of the Olympic Mountains and the Vancouver Island Range, shielding them from the full brunt of incoming Pacific storms. Meteorologists often say the San Juans and Gulf Islands are in the rain shadow of these mountain ranges, and precipitation amounts typically are a third or less of that measured at nearby locations not within the rain shadow.

The same sheltering that protects these islands from the strong coastal winds, seas, and precipitation can also generate locally accelerated winds. The biggest hazards facing skippers cruising through the San Juans and Gulf Islands are funneled winds and the rough seas caused by the interaction of such winds with rapid currents.

During the winter months, southwesterly winds are most common in the San Juans, while southeasterlies are most frequent in the Gulf Islands, particularly from October through March. The variation is due to the northwest–southeast orientation of Vancouver Island. Gale-force winds occur an average of three to four days each month during this season. Occasionally, arctic air blasts out through the Fraser River Valley, creating northeasterly winds that sometimes exceed 60 knots. Such frigid blasts occasionally extend as far west as San Juan and Saturna islands. However, the gale-force northeasterlies usually diminish to 30 knots or less before reaching any of the San Juans or Gulf Islands.

Southwesterly winds are most common through Rosario and Haro straits during the summer, with considerable variability in the San Juans. North of Point Roberts, the prevailing summer wind direction is northwesterly.

The windiest months are also the wettest months: again, primarily October through March. That isn't to say that skippers and their crews can't receive a thorough drenching during the late spring, summer, or early autumn. They can. It's simply far less common.

*The San Juan Islands and Gulf Islands, from Victoria east to Anacortes, north to Vancouver, and west to Nanaimo*

RIVER
FRASER
N
0
2.5
5 MILES
BLAINE HARBOR
TSAWWASSEN
ROBERTS BANK
BELLINGHAM
AND HEADS
BELLINGHAM
BAY
GEORGIA
HALE
PASSAGE
EAST PT.
BELLINGHAM
CHAN.
ORCAS
IS.
EAST SND
BOUNDARY
PASSAGE
GUEMES IS.
THATCHER
PASS
SPIEDEN
CHANNEL
PASSAGE
FRIDAY
HARBOR
LOPEZ IS.
SAN JUAN
IS.
HARO
STRAIT
STRAIT
DECEPTION PASS
WHIDBEY IS.
ROSARIO
VICTORIA

## Sailing Directions:
*The San Juan Islands and Gulf Islands*

| *Weather Hazard* | *Location* | *Specific Caution* |
|---|---|---|
| Strong Winds | East Sound/ Orcas; Rosario, Haro straits; Hale Passage; San Juan and Bellingham channels. Also Saanich Inlet; Trincomali, Stuart, Swanson, and Pylades channels; and Plumper Sound in the Gulfs | Channeling and funneling phenomena typically double prevailing wind speeds in these and similar channels when winds are from the north or south. |
| Strong Winds | Deception Pass; Guemes, Harney and Speiden channels; Thatcher, Obstruction and Peavine passes. Also Satellite Channel, the Fraser River Valley, Active and Porlier passes, and Gabriola Passage in the Gulfs | Channeling and funneling phenomena typically double prevailing wind speeds in these and similar channels when winds are from the east or west. |
| Strong Winds | Fraser River Valley, southern Strait of Georgia | Channeling and funneling phenomena produce cold outflow winds during winter. If high pressure is moving south into B.C. and the pressure gradient reaches or exceeds 10 mb between Hope and Vancouver, gale-force or higher winds are likely. |

## Sailing Directions:

*The San Juan Islands and Gulf Islands*

| *Weather Hazard* | *Location* | *Specific Caution* |
|---|---|---|
| Strong Winds/ Fog | From Victoria east to Anacortes | A surge of moist ocean air and an abrupt shift from northeasterly to southerly or southwesterly winds tend to move rapidly into the area, typically after warm weather from the Willamette Valley north into the San Juans and Gulf Islands. Warning clues include the northward progression of fog/stratus along the Oregon and Washington coasts and a pressure difference of 2 mb or more between Astoria and SeaTac (Seattle), with the higher reading at Astoria. The gusty winds typically move into the San Juans and Gulf Islands from the eastern strait during the afternoon, once these other conditions have been met. Fog or stratus typically develop overnight. |
| Strong Winds/ Rough Seas | Haro, Rosario, and south Georgia straits | A strong sea breeze that develops during fair weather can generate rough seas. Such sea breezes are likely when skies are clear and winds drop to 5 knots or less overnight. The sea breeze develops rapidly in the morning hours. A crossing is most advisable after mid-afternoon, if winds begin diminishing. Once the sea breeze begins to diminish in Georgia Strait, it tends to continue diminishing. |
| Rough Seas | Boundary, Obstruction, and Peavine passes; Pole and Wasp passages | Moderate to strong east winds against a flood tide produce hazardous wave conditions for small craft. |

## Sailing Directions:

*The San Juan Islands and Gulf Islands*

| *Weather Hazard* | *Location* | *Specific Caution* |
|---|---|---|
| Rough Seas | Active and Porlier passes | Moderate to strong northwesterly winds against a flood tide produce hazardous wave conditions just outside the entrances. |
| Rough Seas | Deception Pass | Moderate to strong westerly or near-westerly winds against an ebb tide produce hazardous wave conditions just outside the entrances. |
| Rough Seas | San Juan Channel at Cattle Point | Moderate to strong southerly to southwesterly winds against an ebb tide produce hazardous wave conditions just outside the entrance. |
| Rough Seas | Roberts Bank | Moderate to strong southwesterly winds against ebb currents produce hazardous wave conditions. |
| Fog | Fraser River south, including Bellingham Bay to Anacortes | Dense radiation fog is much more likely than sea fog, particularly during the months of September through February. It is most typical when precipitation is followed by clearing and light winds. |

# Howe Sound to Cape Scott

## Including Northern Georgia Strait, Johnstone Strait, and Queen Charlotte Strait

*Climate Summary*

Considerable variations exist in the general weather patterns of the inside passage between Vancouver Island and the British Columbia mainland. Wind and tide interactions produce most of the weather challenges commonly found in both the Strait of Georgia and Johnstone Strait. Southeasterly winds prevail during the winter from Haro Strait north through Georgia and Johnstone straits into Queen Charlotte Strait. Such winds reach peak velocity when coastal lows move directly across southern Vancouver Island. Velocities typically are 10 to 15 knots stronger in Johnstone Strait than in Queen Charlotte Strait, due to the effect of funneling.

During the summer, northwesterly winds prevail, particularly through Queen Charlotte and Johnstone straits. Sea breezes can produce choppy seas during the day and land or downslope fjord breezes in confined anchorages at night.

In winter, northwesterly winds can pair with strong northwest swell to produce rugged seas through the west entrance of Queen Charlotte Strait. Such conditions typically follow cold frontal passage as high pressure builds over the region. Winter months can also produce arctic outbreaks, where high winds blast through the coastal fjords, bringing cold temperatures, freezing rain, snow, and even ice fog.

Sea fog is a rarity in all but the most northerly section of Queen Charlotte Strait, but radiation fog—as well as precipitation-induced fog—can bring visibility to near-zero during autumn and winter. The few reporting stations available rarely reflect the variety of weather found in this area, requiring skippers to adapt general knowledge to local conditions.

## Sailing Directions:

*Howe Sound to Cape Scott*

| *Weather Hazard* | *Location* | *Specific Caution* |
|---|---|---|
| Strong Winds/ | Howe Sound; Bute, Knight, and Kingcome inlets | Gale-force winds during winter can accelerate through fjords, producing highs winds and rough seas. Freezing rain can occur as the cold arctic air thrusts beneath the warmer, moist air, changing to snow and possibly ice fog. The pressure gradient between Hope and Vancouver can be a helpful indicator, typically exceeding 10 mb just before such outbreaks. |
| Strong Winds | Qualicum Beach, Lasqueti Island | Gale-force winds can occur during summer afternoons/evenings caused by channeling/funneling phenomena. Southwesterly winds moving over Vancouver Island, past Port Alberni, accelerate over Qualicum Beach, often extending as far as Lasqueti Island. |
| Strong Winds/ Rough Seas | Grief Point, Malaspina Strait, Pender Harbour, Welcome Passage, Discovery Passage, Cape Mudge, Goletas Channel, and Baynes Sound | Southeasterly winds accelerate through channeling/funneling, often doubling in velocity. When such winds oppose tidal currents, rough seas develop that are hazardous to small craft. Passages are best planned at, and just after, high-water slack. |
| Strong Winds | Princess Louisa Inlet; Desolation Sound; and Toba, Bute, Knight, Kingcome, and Loughborough inlets | Drainage/gravity winds can occur at night, generally on clear or mostly clear nights, a couple of hours after the sea breeze dies. Such drainage winds aren't usually as strong as the sea breeze, but they do make secure anchoring technique important. |

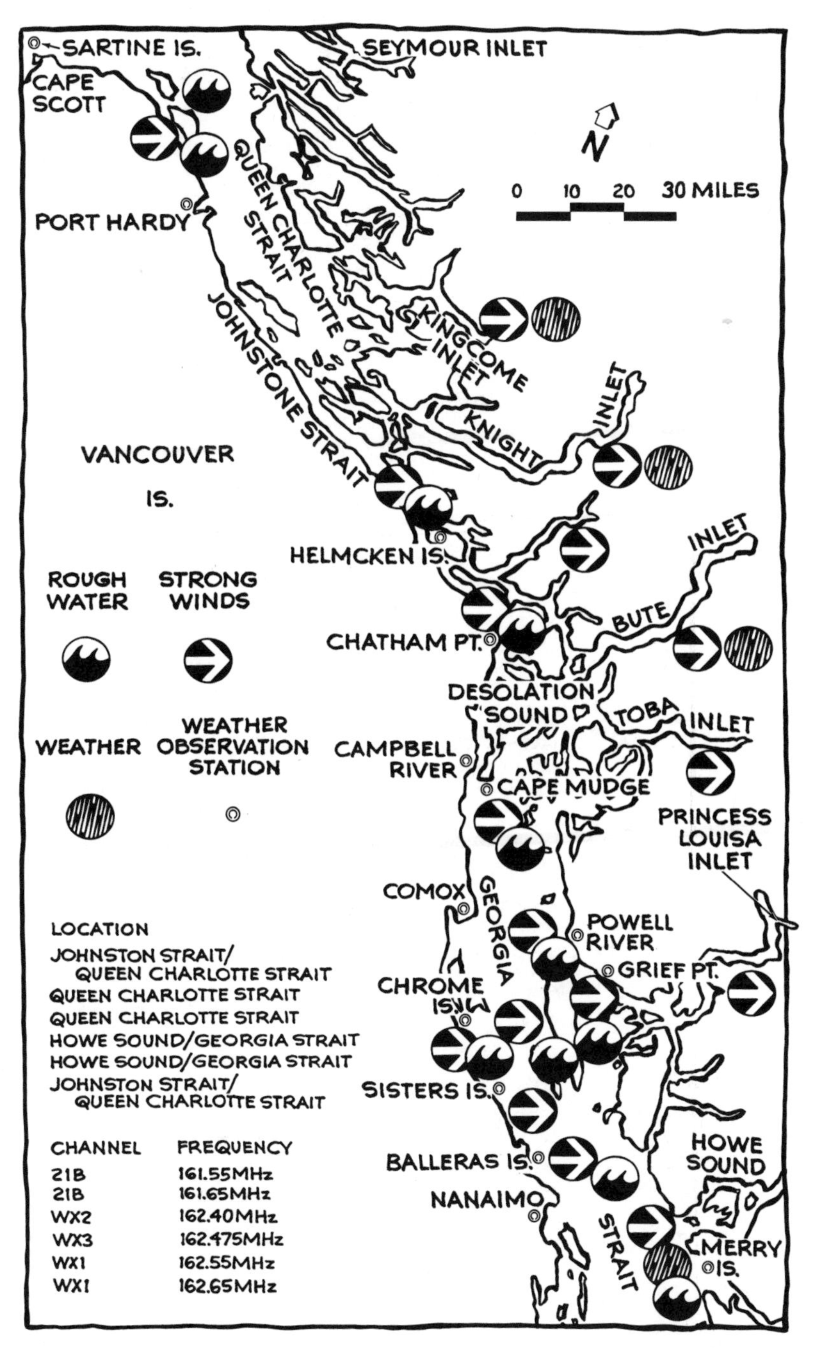


*Howe Sound to Cape Scott, including northern Georgia Strait, Johnstone Strait, and Queen Charlotte Strait*

## Sailing Directions:

*Howe Sound to Cape Scott*

| *Weather Hazard* | *Location* | *Specific Caution* |
|---|---|---|
| Strong Winds/ Rough Seas | Strait of Georgia | Sea breezes, generally from the west, usually reach 25 knots by the early to mid-afternoon hours. Crossings are best delayed until wind velocities begin diminishing in the late afternoon hours. |
| Strong Winds/ Rough Seas | Johnstone Strait | The sea breeze can reach peak velocities of 30 to 35 knots during clear summer days, actually reaching a peak during evening hours. The wind diminishes overnight, making early-morning crossings the best strategy. |
| Rough Seas | Point Atkinson | Tidal currents meet in this area, generating rips and rough seas. Best passage is at slack water. |
| Rough Seas | Kelsey Bay/ Johnstone Strait | Rough water hazardous to small vessels occurs when westerly winds blow against an ebb tide. |
| Rough Seas | West entrance, Queen Charlotte Strait | Heavy ocean swell can pound this area. Check carefully for reports before venturing out. |

# The Pacific Coast of Vancouver Island

*Climate Summary*

Intense lows and their associated strong winds and heavy seas present the biggest weather challenges along Vancouver Island's Pacific coast. Because of the northwest–southeast orientation of Vancouver Island, the strongest winds are typically ahead of lows approaching from the southeast, due to the convergence of wind from the island and over water. The steepness of shoreline terrain determines the strength of such winds. Winter lows producing gale-force winds occur on average at least once a week, although occasionally there's a spell of fine weather lasting two weeks. After cold-front passage, winds usually shift to southwesterlies or northwesterlies. The northwesterlies are typically much stronger.

Outflow winds through gaps in the Vancouver Island Range can produce very strong winds in and near the mouths of associated rivers and sounds, such as Barkley Sound. When a large dome of high pressure exists in the interior of British Columbia, and the cold air is funneling through the Fraser River Valley, be prepared for similar cold outflow winds along the west coast of the island. If a low is approaching, freezing rain may glaze over ships, skippers, and crew alike.

Such strong easterly outflow winds can also develop during the summer, when high pressure brings fair weather. A trough of low pressure builds northward along the Washington coast, extending just offshore from Vancouver Island. Strong offshore winds can rapidly shift to strong onshore winds, along with an abrupt drop in visibility as sea fog surges inland. Summer does produce the longest stretches of light winds, with an entire month passing between gales. But onshore winds, particularly northwesterlies, can easily reach 40 knots.

Such northwesterlies can also generate strong upwelling of cold water along the coast, as it pushes away the warmer surface water. As the cold subsurface water comes in contact with warmer, moist ocean air, dense fog develops rapidly. Visibility tends to be worst along the coast during the late summer months of August and September and best during the spring.

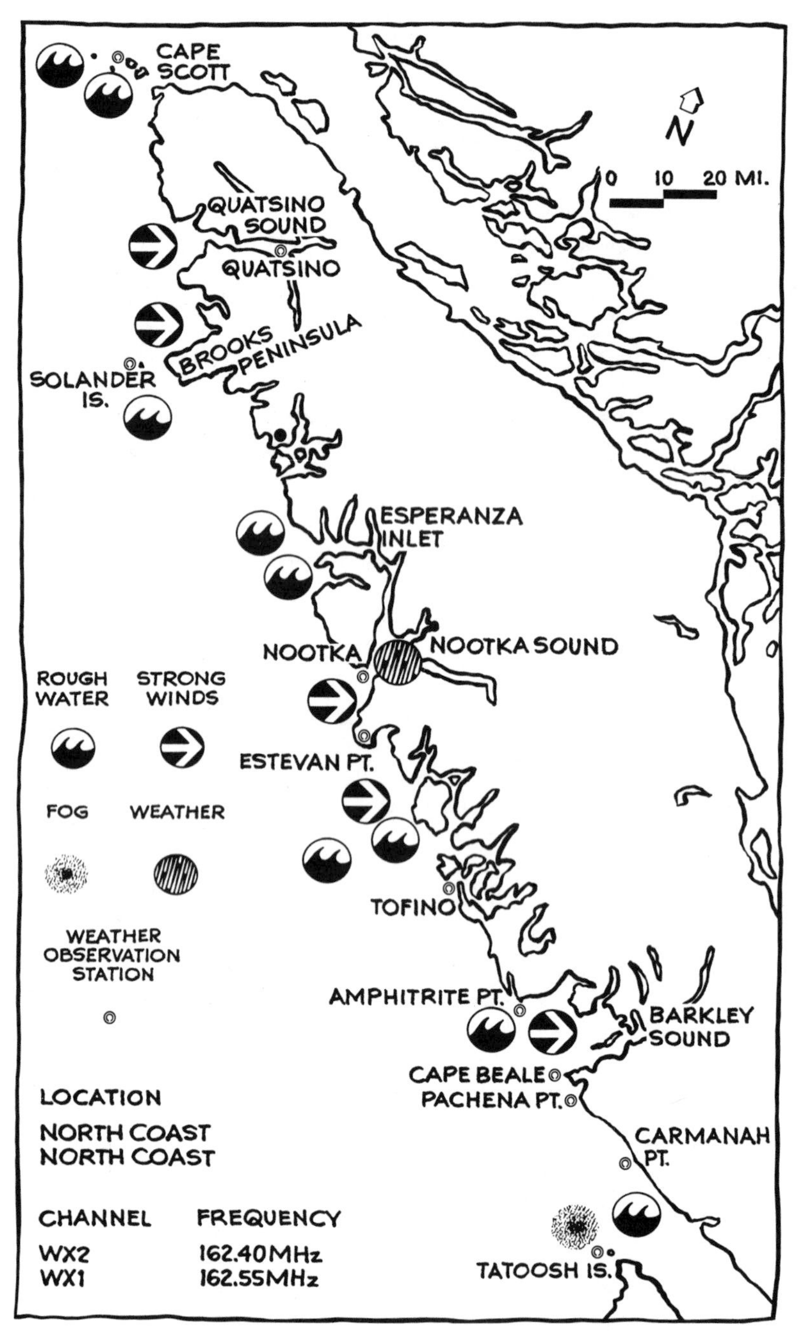


*The Pacific coast of Vancouver Island*

## Sailing Directions:

*The Pacific coast of Vancouver Island*

| *Weather Hazard* | *Location* | *Specific Caution* |
|---|---|---|
| Strong Winds | Brooks Bay | Leeside winds are gusty during gale-force winds from the southeast. |
| Strong Winds | Barkley Sound/ Alberni Inlet | Funneling of southwesterly winds generates higher wind speeds in the inlet, up to double prevailing wind speeds. |
| Strong Winds | Barkley Sound/ Alberni Inlet, Nootka Sound, and Quatsino Sound | High winds occur in winter, when arctic air surges from the Yukon and northern B.C. to the south, pushing westward and accelerating through river valleys cutting through the Vancouver Island Range. Snow, freezing rain, and icing are common in such outbreaks. Watch for reports of easterly outbreaks in Howe Sound and the Fraser River Valley. Pressure differences of 10 mb between Vancouver and Hope (higher pressure) provide a warning sign. |
| Strong Winds/ Rough Water | Brooks Peninsula | When northwesterly or southeasterly winds blow, expect increased wind speeds due to convergence caused by steep mountains along the shore. Wind speeds reach a peak just off Solander Island. Avoid making passage around Brooks Peninsula when tidal currents oppose wind waves or swell. |
| Strong Winds/ Rough Water | Barkley Sound | Ebb tides combine with westerly winds of 25 knots to produce steep, confused seas. |

## Sailing Directions:

*The Pacific coast of Vancouver Island*

| *Weather Hazard* | *Location* | *Specific Caution* |
|---|---|---|
| Strong Winds/ Rough Water | Estevan Point | Southeasterly winds increase due to convergence off the point, combining with strong currents to produce heavy seas. Avoid making passage when winds are strong. Even moderate winds paired with opposing currents or swell can produce hazardous conditions for small craft. |
| Rough Water | Tatchu Point, Nuchatlitz Inlet, Scott Islands, Cook Bank, and Bonilla Point (just offshore) | Westerly waves and swell become steep combined with opposing tidal currents. |
| Rough Water | Scott Channel; also area between Triangle, Sartine, and Lanz Islands | Southwesterly wind waves and swell combine with ebbtide currents to produce steep waves hazardous to small craft, especially when currents reach 3 knots. |
| Rough Water | Vicinity of Nitinat River Bar | Outflow from Nitinat Lake combines with westerly swell, tidal currents, and the bar itself to produce steep waves. |
| Rough Water | Cape Beale/ Trevor Channel | Opposing ebbtide currents and gale-force winds produce hazardous conditions, particularly to small craft. |
| Rough Water | La Perouse Bank/Barkley Sound | Swell and heavy wind waves refracting over the bank generate steep, confused waves near the entrance to Barkley Sound, especially during ebbtide currents. |

## Sailing Directions:

*The Pacific coast of Vancouver Island*

| *Weather Hazard* | *Location* | *Specific Caution* |
|---|---|---|
| Rough Water | Entire coast | Southeast wind-generated waves cross with southwesterly swell to produce steep, confused seas, typical prior to frontal passage. |
| Fog | Barkley Sound | Sea fog is at its worst in this area, particularly during late summer and early autumn months when northwesterly winds have been blowing. |

There's nothing quite as enjoyable as "simply messing about in boats," especially when it's amid the spectacular beauty of the Pacific Northwest. Whether it's the allure of exploring new areas, simply enjoying time with family and friends, or the challenge of racing, there are special places for every taste. There's no question that at times the weather in the Pacific Northwest adds an element of risk to boating. But with the aid of skill, good judgment, and information, each skipper will have many opportunities to cast off for adventure and relaxation with confidence. Here's to fair winds and following seas ... and may our courses cross someday soon!

# Appendixes

## Appendix 1

### Wind-Chill Chart
### (Wind in Miles Per Hour)

| Current Temperature (F) | 0 | 5 | 10 | 15 | 20 | 25 | 30 | 35 | 40 | 45 | 50 |
|---|---|---|---|---|---|---|---|---|---|---|---|
| | 35 | 33 | 21 | 16 | 12 | 7 | 5 | 3 | 1 | 1 | 0 |
| | 30 | 27 | 16 | 11 | 3 | 0 | -2 | -4 | -4 | -6 | -7 |
| | 25 | 21 | 9 | 1 | -4 | -7 | -11 | -13 | -15 | -17 | -17 |
| | 20 | 16 | 2 | -6 | -9 | -15 | -18 | -20 | -22 | -24 | -24 |
| | 15 | 12 | -2 | -11 | -17 | -22 | -26 | -27 | -29 | -31 | -31 |
| | 10 | 7 | -9 | -18 | -24 | -29 | -33 | -35 | -36 | -38 | -38 |
| | 5 | 1 | -15 | -25 | -32 | -37 | -41 | -43 | -45 | -46 | -47 |
| | 0 | -6 | -22 | -33 | -40 | -45 | -49 | -52 | -54 | -54 | -56 |
| | -5 | -11 | -27 | -40 | -46 | -52 | -56 | -60 | -62 | -63 | -63 |
| | -10 | -15 | -31 | -45 | -52 | -58 | -63 | -67 | -69 | -70 | -70 |
| | -15 | -20 | -38 | -51 | -60 | -67 | -70 | -72 | -76 | -78 | -79 |
| | -20 | -26 | -45 | -60 | -68 | -75 | -78 | -83 | -87 | -87 | -88 |
| | -25 | -31 | -52 | -65 | -76 | -83 | -87 | -90 | -94 | -94 | -96 |
| | -30 | -35 | -58 | -70 | -81 | -89 | -94 | -98 | -101 | -101 | -103 |
| | -35 | -41 | -64 | -78 | -88 | -96 | -101 | -105 | -107 | -108 | -110 |

# Appendix 2

## Cloud Identification Chart

*Halo* →
Commonly seen 24–48 hours ahead of precipitation

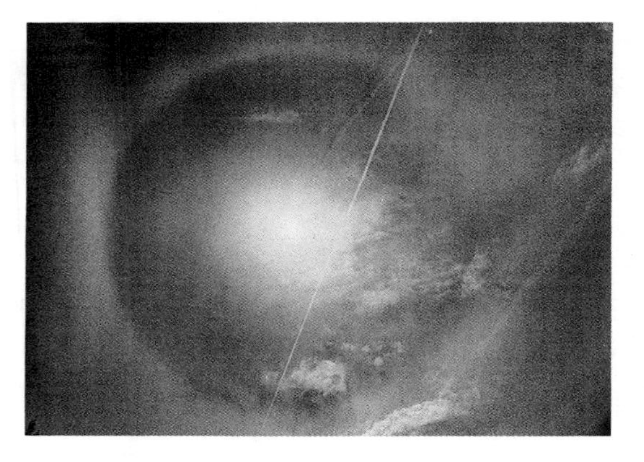

← *Cirrocumulus*
Often changes into cirrus

*Lenticular* →
Wavelike clouds over mountains often suggesting approaching precipitation within 48 hours

← *Stratus*
Layerlike clouds marking approaching warm front, or ocean air

*Cirrostratus* →
Often indicates approaching warm front

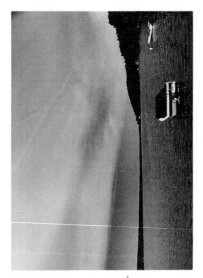

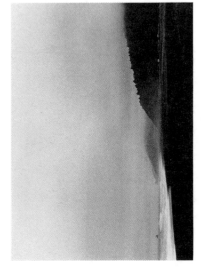

← *Altostratus*
When part of approaching warm front, follows cirrostratus

*Nimbostratus* →
Stratus clouds producing widespread precipitation and low visibility

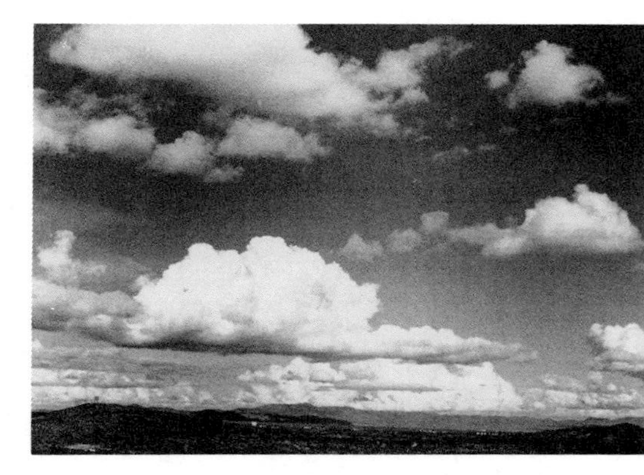

← *Cumulus*
With continued upward growth these suggest showers later in the day

*Altocumulus* →
High-based clouds often indicating potential for thunder, rain showers

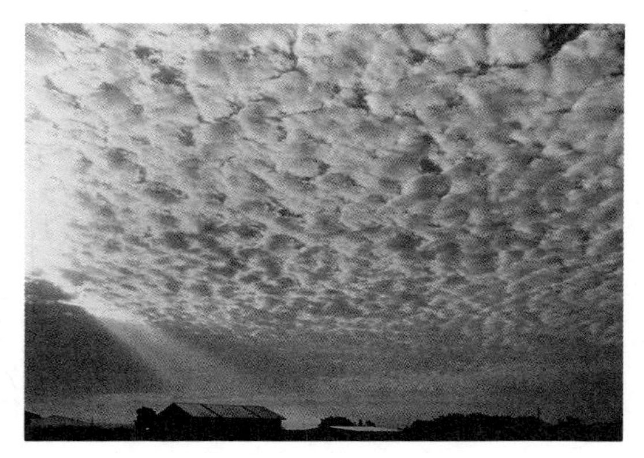

← *Cumulonimbus*
Cumulus producing rain, snow, or thunder and lightning

*Stratocumulus* →
Lumpy, layered clouds often following a cold front, suggesting showers

# Appendix 3

## Useful Equivalents and Conversions

| *Temperature (Celsius)* | *Temperature (Fahrenheit)* | |
|---|---|---|
| 0 | 32.0 | (Freezing Point of Water) |
| 10 | 50 | |
| 20 | 68 | |
| 30 | 86 | |
| 37 | 98.6 | (Normal Body Temperature) |
| 40 | 104 | (Heat-Wave Conditions) |
| 100 | 212 | (Boiling Point of Water) |

*Actual Conversion from Fahrenheit to Celsius*

(Degrees Fahrenheit–32) x 5/9
OR
(Degrees Fahrenheit–32) x .556

*Actual Conversion from Celsius to Fahrenheit*

(9/5 x Degrees Celsius) + 32
OR
(1.8 x Degrees Celsius) + 32

*Equivalent Marine Depth and Distance Measurements*

| | |
|---|---|
| 6 Feet | 1 Fathom |
| 5,280 Feet | 1 Statute Mile |
| 6,076 Feet | 1 Nautical Mile |
| 1 Statute Mile | .869 Nautical Mile |
| 1 Nautical Mile | 1.151 Statute Miles |
| 1 Degree Latitude | 60 Nautical Miles |
| 1 Minute Latitude | 1 Nautical Mile |
| 1 Meter | 3.3 Feet |
| 1 Foot | .30 Meters |

## *Atmospheric Pressure*

| | |
|---|---|
| 1 Inch of Mercury | 33.8 Millibars |
| 1 Millibar | .03 Inches of Mercury |

## *Wind Speeds*

| | |
|---|---|
| 1 Knot | 1.15 Statute Miles per Hour |
| 1 Statute Mile per Hour | .87 Knot |
| 1 Knot | 1.8 Kilometers per Hour |
| 1 Kilometer per Hour | .6 Miles per Hour |
| 1 Kilometer per Hour | .55 Knots |
| 1 Mile per Hour | 1.6 Kilometers per Hour |

# Appendix 4

## Electronic Data Sources

### Weather Data by Computer

CompuServe
P.O. Box 20212
5000 Arlington Center Boulevard
Columbus, OH 43220
(800) 848-8199
(614) 457-8650

Global Weather Dynamics
2400 Garden Road
Monterey, CA 93940
(800) 538-9507
(408) 649-4500

Ocean Routes/Tymshare
Weather Network Division
680 West Maude Avenue
Sunnyvale, CA 94086-3518
(408) 245-3600

Prodigy
445 Hamilton Avenue
White Plains, NY 10601
(800) 284-5933

WSI Corporation
41 North Road
Bedford, MA 01730-0902
(617) 275-5300

### Weather Data by Fax

ZFAX/Zephyr
40 Washington Street
Westborough, MA 01581-0500
(800) 876-1232

JeppFax
Jeppesen
55 Inverness Drive East
Englewood, CO 80112
(800) 621-5377

# Appendix 5

## Suggested References

Bishop, Dr. Joseph M., *A Mariner's Guide to Radiofacsimile Weather Charts.* Westborough, Massachusetts: Alden Electronics, 1991.

Canada Department of Fisheries and Oceans, Scientific Information and Publications Branch. *Sailing Directions, British Columbia Coast.* Ottawa, 1989.

Environment Canada, Atmospheric Environmental Service. *Marine Weather Hazards Manual.* Ministry of Supply and Services, No. 56-76, 1990.

Gedzelman, Stanley D., *The Science and Wonders of the Atmosphere.* New York: John Wiley & Sons, 1980.

Kawaky, Joseph, *Capn. Jack's Tide and Current Almanac—British Columbia.* Port Ludlow, Wash.: Marine Trade Publications, 1993 (updated yearly).

Kawaky, Joseph, *Capn. Jack's Tide and Current Almanac—Puget Sound.* Port Ludlow, Wash.: Marine Trade Publications, 1993 (updated yearly).

Kotsch, William J., *Weather for the Mariner.* Annapolis: Naval Institute Press, 1983.

Thomson, Richard E., *Oceanography of the British Columbia Coast.* Ottawa: Department of Fisheries and Oceans, 1981.

U.S. Department of Commerce, National Oceanic and Atmospheric Administration. *U.S. Coast Pilot: Pacific Coast.*

U.S. Department of Transportation, Federal Aviation Administration. *Aviation Weather.* Advisory Circular No. 00-61, 1987.

U.S. Department of Transportation, Federal Aviation Administration. *Aviation Weather Services.* Advisory Circular No. 00-45C, 1987.

# Index

## D



## E



## F



## G



## H



## I



## J

## R



## S



## T



## U



## V

## W



## Y

**ABOUT THE AUTHOR:** Jeff Renner has enjoyed and experienced the marine environment of the Pacific Northwest from all perspectives: from above as a commercial seaplane pilot, upon the water as a sailor and kayaker, and beneath the surface as a scuba diver. Fascinated from childhood by the underwater adventures of Jacques Cousteau, Jeff has explored above and below the world's waters from the Caribbean to the South Pacific, and along the Pacific Coast from Mexico to Alaska. First-hand encounters with challenging marine weather, together with a degree in atmospheric sciences from the University of Washington and more than a decade of experience as a professional forecaster, have helped Jeff develop workable strategies for anticipating and coping with weather afloat. As meteorologist for KING-TV in Seattle, he endeavors to bring that experience and dedication to improving the quality and scope of weather information for all who work or play on the waters of the Pacific Northwest.

**THE MOUNTAINEERS,** founded in 1906, is a nonprofit outdoor activity and conservation club, whose mission is "to explore, study, preserve, and enjoy the natural beauty of the outdoors...." Based in Seattle, Washington, the club is now the third-largest such organization in the United States, with 14,000 members and four branches throughout Washington State.

The Mountaineers sponsors both classes and year-round outdoor activities in the Pacific Northwest, which include hiking, mountain climbing, ski-touring, snowshoeing, bicycling, camping, kayaking and canoeing, nature study, sailing, and adventure travel. The club's conservation division supports environmental causes through educational activities, sponsoring legislation, and presenting informational programs. All club activities are led by skilled, experienced volunteers, who are dedicated to promoting safe and responsible enjoyment and preservation of the outdoors.

The Mountaineers Books, an active, nonprofit publishing program of the club, produces guidebooks, instructional texts, historical works, natural history guides, and works on environmental conservation. All books produced by The Mountaineers are aimed at fulfilling the club's mission.

If you would like to participate in these organized outdoor activities or the club's programs, consider a membership in The Mountaineers. For information and an application, write or call The Mountaineers, Club Headquarters, 300 Third Avenue West, Seattle, Washington 98119; (206) 284-6310.